PREDICTING THE REMAINING ECONOMIC LIFE OF WASTEWATER PIPES

PHASE I: DEVELOPMENT OF STANDARD DATA STRUCTURE TO SUPPORT WASTEWATER PIPE CONDITION AND PERFORMANCE PREDICTION

Sunil K. Sinha, Ph.D.
Thiti Angkasuwansiri
Virginia Tech

Richard Thomasson
Malcolm Pirnie

2010

WERF

The Water Environment Research Foundation, a not-for-profit organization, funds and manages water quality research for its subscribers through a diverse public-private partnership between municipal utilities, corporations, academia, industry, and the federal government. WERF subscribers include municipal and regional water and wastewater utilities, industrial corporations, environmental engineering firms, and others that share a commitment to cost-effective water quality solutions. WERF is dedicated to advancing science and technology addressing water quality issues as they impact water resources, the atmosphere, the lands, and quality of life.

For more information, contact:
Water Environment Research Foundation
635 Slaters Lane, Suite G-110
Alexandria, VA 22314-1177
Tel: (571) 384-2100
Fax: (703) 299-0742
www.werf.org
werf@werf.org

This report was co-published by the following organization. For non-subscriber sales information, contact:

IWA Publishing
Alliance House, 12 Caxton Street
London SW1H 0QS, United Kingdom
Tel: +44 (0) 20 7654 5500
Fax: +44 (0) 20 7654 5555
www.iwapublishing.com
publications@iwap.co.uk

Library of Congress Catalog Card Number: 2009928961
IWAP ISBN: 978-1-84339-333-7/1-84339-333-6

Virginia Tech

The research on which this report is based was developed, in part, by the United States Environmental Protection Agency (EPA) through Cooperative Agreement No. CR-83155901 with the Water Environment Research Foundation (WERF). However, the views expressed in this document are not necessarily those of the EPA and EPA does not endorse any products or commercial services mentioned in this publication. This report is a publication of WERF, not EPA. Funds awarded under the Cooperative Agreement cited above were not used for editorial services, reproduction, printing, or distribution.

ACKNOWLEDGEMENTS

The authors are grateful for the support from the Water Environmental Research Foundation. The authors also acknowledge the contribution of various utilities, national pipe associations, U.S. Environmental Protection Agency, and international organizations for providing their comments and feedback in an effort to successfully conduct this research project.

Research Team

Principal Investigator:
Sunil K. Sinha, Ph.D.
Virginia Tech

Project Team:
Thiti Angkasuwansiri
Virginia Tech

Richard O. Thomasson
Malcolm Pirnie

WERF Project Subcommittee

Terry Martin
Seattle Public Utilities

Kevin Hadden
Orange County Sanitation District

Greg Cawston
Greg Kane
Sydney Water Corporation

Sanjay Patel
MWRDGC

Wayne Green
Green Management Ltd.

Kendall Jacob, P.E.
Cobb County Water System

Tim Madhanagopal
Orange County Utilities Department

Ted Regan
Massachusetts Water Resources Authority (MWRA)

Steve Krai
Los Angeles County Sanitation District

Maureen Hodgins
Water Research Foundation

Chris Royce
Steve Kaye
Anglian Water Services, Ltd.

Daniel Murray
U.S. Environmental Protection Agency

Stephen Whipp
United Utilities North West (UUNW)

Ron Booth
Westin Engineering, Inc.

James Thomson, P.E.
Jason Consultants LLC, Inc.

Participating Utilities

City of Atlanta, Atlanta, GA – John Griffins
Seattle Public Utility, Seattle, WA – Terry Martin
Orange County Sanitation District, Fountain Valley, CA – Kevin Hadden
Pittsburgh Water and Sewer Authority, Pittsburgh, PA – Michael Lichte
VPI Sanitation Authority, Blacksburg, VA – Ray Smoot, Michael Vaught, Katherine Smith

National Pipe Association Participants

American Concrete Pipe Association – Kim Spahn
National Clay Pipe Institute – Michael Van Dine
UniBell PVC Pipe Association – Robert Walker
Ductile Iron Pipe Research Association – Gregg Horn

United States Environmental Protection Agency

Michael Royer
Dan Murray
Dennis Lai
Richard Field
Ari Selvakumar

International Organizations

National Research Council, Canada, Dana Vanier
CSIRO, Australia, Stewart Burn
Kirby Associates, UK, Richard Kirby

Water Environment Research Foundation Staff

Director of Research: Daniel M. Woltering, Ph.D.
Program Director: Walter L. Graf, Jr.

ABSTRACT AND BENEFITS

Abstract:

Accurate prediction of wastewater pipe structural and functional deterioration plays an essential role in the utility asset management process and capital investment planning. The key to implementing an asset management strategy is a comprehensive understanding and prediction of asset condition and performance. The primary objective of this research is therefore to develop protocols and methods for predicting the remaining economic life of wastewater pipe assets. The limits of deterioration prediction capabilities are not in mathematical models or statistical analysis methods, but in lack of accurate and consistent data. This report presented the short-term Phase 1, which was completed with results from intensive literature reviews, various interviews with utilities, and pipe associations. In this phase, the research team investigated the life cycle of wastewater pipeline and identified the causes of pipe failure in different phases including design, manufacture, construction, operation, and maintenance and repair, rehabilitation, and replacement. The research team prepared various modes and mechanisms of pipe failure in wastewater infrastructure system as well as identified environmental and societal consequences of the failure. After reviewing all relevant reports and utility databases, the research team developed a set of standard pipe parameters (data structure) and pipe data collection methods. The data structure is classified into Gold, Silver, Bronze, and Wood standards.

Benefits:

- Presents information about wastewater pipe failure modes and mechanisms as well as life cycle during various phases including design, manufacture, construction, operation and maintenance, and repair/rehabilitation/replacement.
- Summarizes wastewater pipe failure consequences and complexity of repair/rehabilitation/replacement.
- Provides standard data structure and data collection methods that would enable effective and systematic data collection, and data storage for establishing asset management programs and analyzing the performance of wastewater pipe infrastructure systems.
- Offers utilities, who manage wastewater pipe databases, five case studies that describe the overview of each city wastewater system, utility data collection, and utilities' GIS capabilities.

Keywords: Asset management, wastewater pipe infrastructure, standard data structure, pipe failure mode and mechanism, structural and functional deterioration.

TABLE OF CONTENTS

LIST OF TABLES

LIST OF FIGURES

LIST OF APPENDICES

LIST OF ACRONYMS AND ABBREVIATIONS

AC	asbestos cement
APUB	Atlanta Public Utilities Board
ArcSDE	advanced spatial data server
CARE-S	computer aided rehabilitation of sewer networks
CCTV	closed circuit television
CMOM	capacity, management, operation and maintenance
CRI	condition rating index
CSO	combined sewer overflow
FTP	file transfer protocol
GIS	geographic information system
H_2S	hydrogen sulfide
ICG	internal condition grade
mgd	million gallon per day
MH	maintenance holes
NAD	North American Datum
NASSCO	National Association of Sewer Service Companies
OCSD	Orange County Sanitation District
PACP	Pipeline Assessment and Certification Program
PVC	polyvinyl chloride
PWSA	Pittsburgh Water and Sewer Authority
RCP	reinforced concrete pipe
SAM	strategic asset management
SDSC	San Diego Super Computer Center
SPU	Seattle Public Utilities
SSES	Sewer System Evaluation Survey
SWIMS	Sustainable Water Infrastructure Management System
UCSD	University of California, San Diego
UKWIR	United Kingdom Water Industry Research
U.S. EPA	United States Environmental Protection Agency
WEF	Water Environmental Federation
WERF	Water Environmental Research Foundation
WRc	Water Research Center

EXECUTIVE SUMMARY

Water, particularly drinking water and wastewater, plays a critical role in every aspect of civilization: agriculture, industry, economy, environment, recreation, transportation, culture, and health. Although clean water and sanitary conditions remain an elusive luxury in many parts of the world, we, as Americans, take them for granted. Much of America's drinking water and wastewater infrastructure; however, is old and deteriorating. A crisis looms as our demands on these systems increase. The costs associated with renewal of these aging systems are staggering. There is a critical disconnect between the methodological remedies for infrastructure rehabilitation/renewal/replacement problems and the current sequential or isolated manner of rehabilitation/renewal analysis and execution. These disconnects manifest in the need for a holistic systems perspective to address the rehabilitation and renewal problem. New tools are needed to provide intellectual support for water infrastructure decisions. Such decisions are necessary to sustain economic growth, environmental quality, and improve societal health.

Developing an efficient infrastructure management system can provide fast and reliable decision-making tools. These tools are needed to handle the large volume of deteriorating buried pipeline infrastructure systems, particularly wastewater pipelines, which pose serious threats to the human health and the environment if they fail.

Condition and performance assessment and deterioration modeling are rapidly becoming an increasing part of life-cycle asset management activities in the United States. In this research, a total-system approach was taken for development of protocols and methods for predicting the condition/performance deterioration of wastewater pipe infrastructure assets. The research project is composed of three major tasks:

1. **Short term (Phase 1, August 2007-September 2008):** Collect all relevant data, tools, and methodologies to evaluate and plan for pipe infrastructure asset end-of-life predictions in an overall utility asset management program;
2. **Medium term (Phase 2 August 2008-September 2009):** Conduct analysis, assessment, and integration of data/information with existing and new methodologies for development of standard condition rating scale;
3. **Long term (Phase 3, August 2009-September 2010):** Develop robust probabilistic deterioration models for predicting the remaining economic life of wastewater pipe assets.

This report details the short-term phase, which was completed with results from intensive literature reviews, and various interviews with utilities and pipe associations. In this phase, the research team investigated the life cycle of wastewater pipelines and identified the causes of pipe failure in different phases including design, manufacture, construction, operation and maintenance, and repair/rehabilitation/replacement. The research team prepared various modes and mechanisms of pipe failure in wastewater infrastructure systems. Environmental and societal consequences of wastewater pipe failure are identified as well.

The research team at Virginia Tech extensively reviewed reports produced by Water Environment Research Foundation (WERF), American Water Works Association Research Foundation (AwwaRF), U.S. Environmental Protection Agency (U.S. EPA), UK Water Industry Research (UKWIR), Computer Aided Rehabilitation of Water Networks (CARE-W), Computer Aided Rehabilitation of Sewer Networks (CARE-S), National Research Council Canada, and others. Based on their research, the research team developed a set of standard pipe parameter list (data structure). The data structure is classified into Gold, Silver, Bronze, and Wood standards.

In addition to the data structure, the research team developed a systematic data collection methodology that enables utilities to efficiently handle data collection and data storage. Utilities and researchers who have been collecting the wastewater pipe infrastructure data can use these data structure and collection methods as a standard for an asset management program and analyzing the performance of wastewater pipe infrastructure system.

This report addresses, in detail, short-term tasks for development of protocols and methods for predicting the remaining economic life of wastewater pipe infrastructure assets. This research is a part of WERF's Strategic Asset Management Challenge.

CHAPTER 1.0

INTRODUCTION

Chapter Highlights

This chapter presents the introductory materials related to pipe infrastructure system:

- **Overview:** Today, municipal governments are facing an infrastructure crisis requiring costly renewal beyond their capacity. There has been a steady decline in the state of our water infrastructure over the past two decades and a growing concern is that these facilities may be inadequate both for current requirements and projected future growth.

- **Funding:** Funding for wastewater systems needs is limited.

- **Decision Supporting Tools:** There is a critical disconnect between the remedies for infrastructure renewal problems and the current sequential or isolated manner of renewal analysis and execution. New tools are needed to provide the intellectual support for utility asset decisions necessary to sustain economic growth, environmental quality, and improved societal health.

- **Research:** This report presents the following research steps:

    1. Standard data model to support the model to predict remaining economic life of wastewater pipe (Phase 1, August 2007-September 2008)
    2. Analysis, assessment, and integration of standard data model for development of uniform pipe condition rating scale (Phase 2, August 2008-September 2009)
    - Robust probabilistic deterioration models for predicting the remaining economic life of wastewater pipe (Phase 3, August 2009-September 2010)

According to the Water Infrastructure Network, "new solutions are needed to what amounts to nearly a trillion dollars in critical water and wastewater investments over the next two decades. Not meeting the investment needs of the next 20 years risks reversing the public health, environmental, and economic gains of the last three decades." [1]. Most cities and towns started building drinking water and collection systems more than 100 years ago and many of these systems have not received adequate upgrades, maintenance, repair, and rehabilitation over time [2]. Today, municipal governments are facing an infrastructure crisis requiring costly renewal beyond their capacity. There has been a steady decline in the state of our water infrastructure over the past two decades and there is a growing concern is that these facilities may be inadequate both for current requirements and projected future growth [2]. Funding for these needs is limited, and a deferred maintenance, out-of-sight, out-of-mind philosophy still prevails in many regions. Recently the American Society of Civil Engineering in its 2005 assessment of the nation's infrastructure assigned the grade "D" and estimated the five-year investment needs to be in excess of $1.3 trillion [3]. It is estimated that the cost of replacing all water mains in the United States would run to $348 billion [4]. Although the federal government has spent more than $71 billion on wastewater treatment programs since 1973, the nation's 19,000 wastewater systems still face enormous infrastructure funding needs in the next 20 years to replace pipes and other constructed facilities that have exceeded their design life [5]. With billions of dollars being spent yearly for water infrastructure, the systems face a shortfall of at least $21 billion annually to replace aging facilities and comply with existing and future federal regulations [1].

Monetary investment alone will not resolve this dilemma; it must be met with a new approach to sustainable water infrastructure engineering and management. There is a critical disconnect between the methodological remedies for infrastructure renewal problems and the current sequential or isolated manner of renewal analysis and execution. These disconnect manifests in the need for a holistic systems perspective. New tools are needed to provide the intellectual support for utility assets decisions necessary to sustain economic growth, environmental quality, and improved societal health.

Developing an efficient infrastructure asset management system can provide fast and reliable decision making tools needed to handle the large volume of deteriorating buried pipeline systems, particularly wastewater pipelines, which pose serious threats to the environment if they fail. A key to implementing asset management strategies is a comprehensive understanding of asset condition and performance. Wastewater utilities need protocols and methods for predicting the expected remaining economic life of utility assets. Predictive modeling can be used to provide a time stream of benefits, costs, and level of service for both single and groups of assets.

Condition/performance assessment and deterioration modeling are rapidly becoming an increasing part of life-cycle asset management activities in the United States. In this research a "total-systems" approach has been taken for development of protocols and methods for predicting the condition/performance deterioration of wastewater pipe asset. This goal requires a short-, medium-, and long-term phase approach.

The proposed research will focus on collecting all relevant data, tools, and methods to evaluate and plan for wastewater pipe infrastructure asset end-of-life predictions in an overall utility asset management program (Phase 1, short-term), conducting analysis, assessment, and integration of data/information with existing methodologies for development of standard condition rating scale (Phase 2, medium-term), developing robust probabilistic deterioration models for predicting the remaining economic life of wastewater pipe assets (Phase 3, long-term). The research will evaluate this approach in a real-world pilot project. The proposed research program is a model system that can be implemented at other municipalities and utilities around the country. The overall objective of this research is to provide utilities with tools and methods integrate into their asset management programs.

1.1 Summary Description of WERF Research Tracks

What follows is a brief summary of the key tasks involved in each track. Detailed tasks and deliverables are articulated in the proposals of the two selected research teams—EMA-GHD-MWH Consortium and Virginia Tech—and will be spelled out in individual task orders:

Track 1. Stakeholders Communication Framework and Tools (EMA Team)

This research track encompassed internal and external stakeholder communication about a utility's asset management drivers and their overall strategic asset management (SAM) program. It is focused on two major tasks:

1. Establish current perceptions of SAM among elected/appointed officials.
2. Develop communication tools to alter perceptions and overcome barriers as necessary.

Before wastewater and water utilities can fully adopt and successfully implement a SAM program, decision makers such as elected and appointed officials and other stakeholders need to clearly understand the benefits, costs, and value strategic asset management offers a community. The research team's approach built on work already accomplished by other public education programs on asset management. Working closely with key public sector partners – the U.S. Conference of Mayors and International City/County Management Association – the team conducted focus groups and surveys to determine perceptions and identify barriers.

Track 2. SAM Benchmarking and Case Studies (EMA Team)

This research track embraced activities aimed at developing a benchmarking framework for SAM to assist utilities in benefiting from "best practices" that cover the technical aspects of SAM as well as organizational, technology, and communication practices. The benchmarking approach will determine the areas of study of "best practices" that provide the most benefit for utilities. Potential benchmarking partners (both national and international) were identified and invited to participate in the screening benchmarking process. The research team evaluated the SAM processes and practices of the selected utilities to develop case studies and lessons learned in the application of asset management principles and best or leading-edge practices. Detailed evaluations of practices were made based on structured site visits. Leading strategic asset management practices were documented.

Track 3. Decision Support and Implementation Guidance (EMA Team)

This track was organized around two major elements:

1. Analytic/decision support tools – Development of staged user-friendly analytic and decision support tools (gap analysis, cost-benefit analysis, risk management, life-cycle costing, condition assessment, etc.) for SAM implementation and continuous improvement.
2. Implementation – Provide practical, accessible guidance for implementation of an asset management program in incremental stages to meet the needs of a wide range of utilities.

Available SAM tools were inventoried and reviewed. Sources include agency practitioners, academics, research organizations, and professional associations as well as other software developers. The team will also drew on internal databases of known products and systematically searched for leads on available tools, especially through Internet searches. Within the cost constraints, new tools will be developed or existing tools will be leveraged to suit requirements. The initial suite of key tools developed included: Gap analysis tool (SAMGAP, with WaterRF co-funding), Benefit-Cost Analysis tool to aid decision making (with the Global Water Research Coalition (GWRC)/ UKWIR/WaterRF co-funding), Risk Management Tool (with GWRC/UKWIR, WaterRF, Water Service Authorities of Australia (WSAA) co-funding). A condition assessment selection tool (based on a condition assessment strategies and protocols report by WERF) is planned. Additional tools will be prioritized as the SAM Challenge proceeds.

Track 4. Prediction of the Remaining Asset Life (EMA Team and Virginia Tech Team)

Track 4 investigated and developed techniques, tools, and methods for predicting the remaining economic life of utility assets (both buried and aboveground). The EMA team studied the aboveground assets and contributed information on international methodologies and approaches for buried assets. The Virginia Tech team investigated the buried assets. In the short-term, the two research teams defined the state of practice on prediction of performance/decay curves via a literature and tool study. The Virginia Tech approach also involves data mining of detailed asset information on the buried assets, which are more complex to predict than the aboveground assets. In addition to structural and environmental factors, the Virginia Tech team will also study the impact of operational/functional and economic factors on pipe failure. It is intended to develop a standard procedure for performance/condition rating and the study will employ advanced artificial intelligence techniques and/or mathematical modeling methodologies for deterioration models. The investigation and development of techniques, tools and methodologies for the definition of the remaining economic life of utility assets will be undertaken in such a way as to provide a stepwise development of concepts and protocols. Track 4 will also develop methodologies for setting up and maintaining a national/global database for performance/decay curves and condition assessment correlations.

Track 5. Internal Communications Management and SIMPLE (Sustainable Infrastructure Management Program Learning Environment) Knowledge Base Enhancement

This track has two key activities. The first is to establish a secure, dedicated website (Team Room) to be used by the research team personnel for managing internal communications regarding the content and progress of various deliverables under the Challenge. (Note: WERF periodically draws information from the Team Room website, but maintains its own dedicated website for external communications with the broader stakeholders. The second activity ensures technical compatibility of the contents or the outputs (documents, tools, etc.) of the SAM Challenge to enable their integration into SIMPLE. The described WERF research tracks are graphically presented in Figure 1-1.

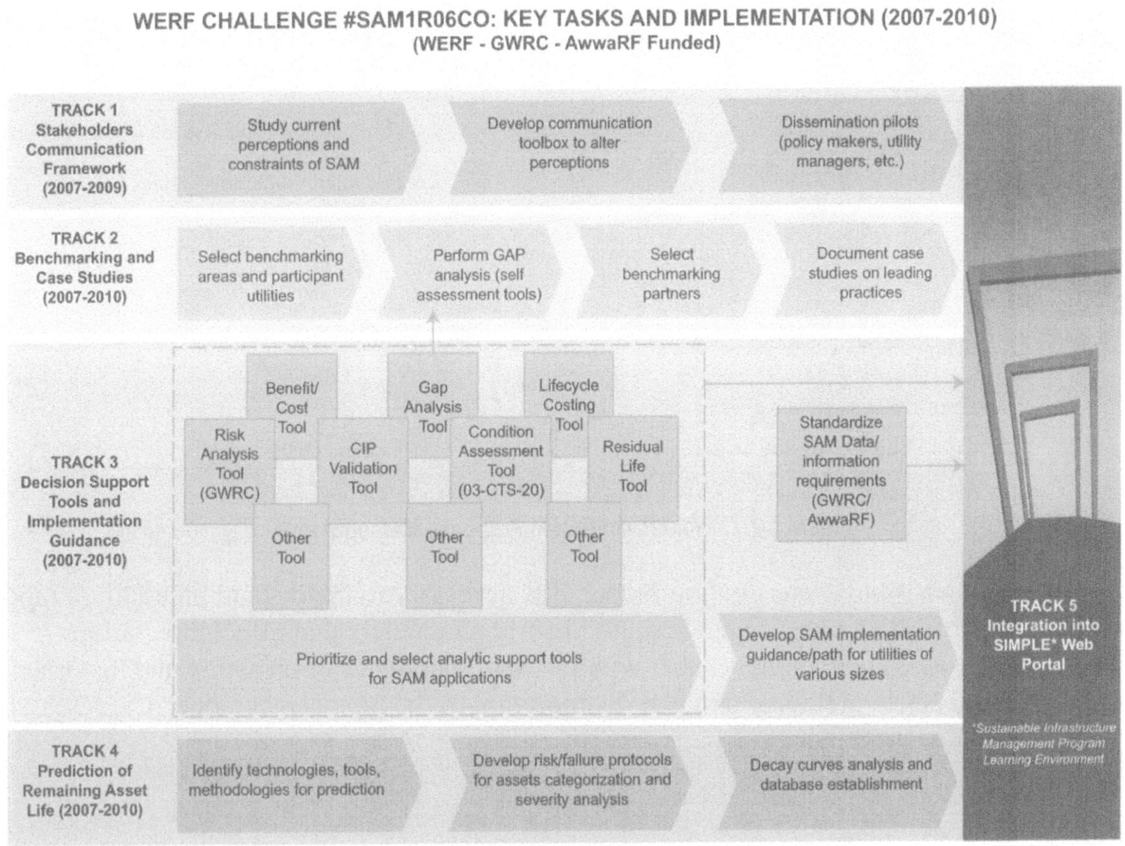

Figure 1-1. WERF Research Tracks.

1.2 Track 4 Virginia Tech Research Progress

Presently, utilities rely on mostly CCTV (closed-circuit television) in evaluating pipes and making rehabilitation decisions. The CCTV inspection is rated using in-house or National Association of Sewer Service Companies (NASSCO)'s rating systems. However, the CCTV doesn't capture all data needed in predicting wastewater pipe failure. This research proposed methods to evaluate the pipe condition by using not only CCTV data, but also structural, environmental, operational, and other data. Based on these data, a better condition index and prediction model can be developed. Figure 1-2 shows the current practice employed by many utilities compared to the proposed methodology.

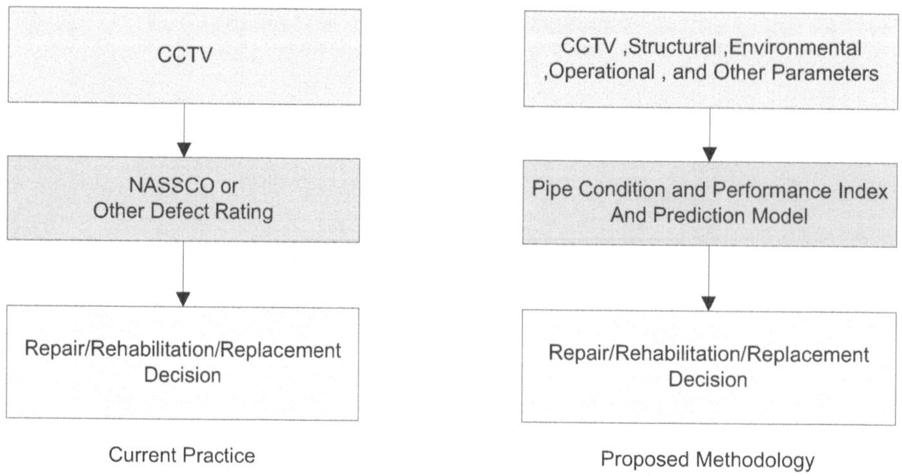

Figure 1-2. Current Practice and Proposed Methodology

To develop a robust deterioration model, it is necessary to understand pipe failure modes and mechanisms based on pipe life cycles, and identify parameters that affect pipe failure processes. The research team believes that with the right parameter information and historical performance, analysis to predict and model deterioration of the pipe can be done. The research approach is illustrated in Figure 1-3.

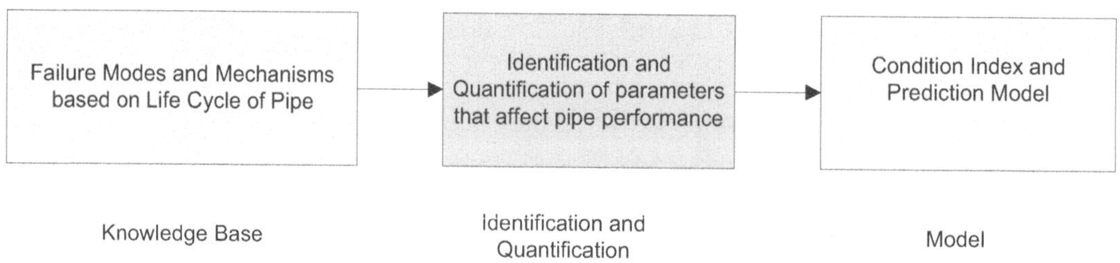

Figure 1-3. Research Approach.

The short-term Phase 1 was completed with results from literature reviews, various interviews with utilities, and pipe associations. In this phase, the research team investigated the life cycle of wastewater pipeline and identified the causes of pipe failure in different phases including design, manufacture, construction, operation and maintenance, and repair/rehabilitation/replacement. The research team prepared various modes and mechanisms of pipe failure in wastewater infrastructure systems. Failure modes of wastewater pipes are different type of failures found in the pipe. Failure mechanism is a condition that causes the pipe to reach one or combined strength and serviceability limit states. Environmental and societal consequences from wastewater pipe failure are identified as well.

Moreover, the research team suggested national standard wastewater data structure and wastewater data collection methodologies that would enable effective and systematic data collection and data storage. Utilities and researchers who collect the wastewater pipe infrastructure data can use these data structure and data collection methodologies as a standard for their asset management program or studying wastewater pipe infrastructure systems.

The proposed standard data structure will help develop a standard condition rating system for pipes (future research work under Phase 2), which will form a fundamental basis for development of a pipe performance prediction model (future research work under Phase 3).

The future research work proposed here can be built upon ongoing research by the principal investigator, WERF, and other organizations. Through collaboration with other utilities/organizations [U.S. EPA; Department of Environmental Protection; Rural Utilities Service; AWWA ; American Public Works Association; American Society of Civil Engineers (ASCE),; Buried Asset Management Institute-International (BAMI-I); Commonwealth Scientific and Industrial Research Organization (CSIRO), Australia; The Foundation for Scientific and Industrial Research (SINTEF), Norway; Dresden University of Technology, Germany UKWIR, United Kingdom; National Research Council (NRC) Canada; and 17 cities in the United States] and by building on existing research [SWIMS, CARE-S , Computer Aided Rehabilitation of Water Networks (CARE-W), WERF, and other research projects], the research outcome will have a wide appeal and benefits to utilities.

Virginia Tech Research Team

Figure 1-4 shows the project organizational chart for Phases 2 and 3 of this project.

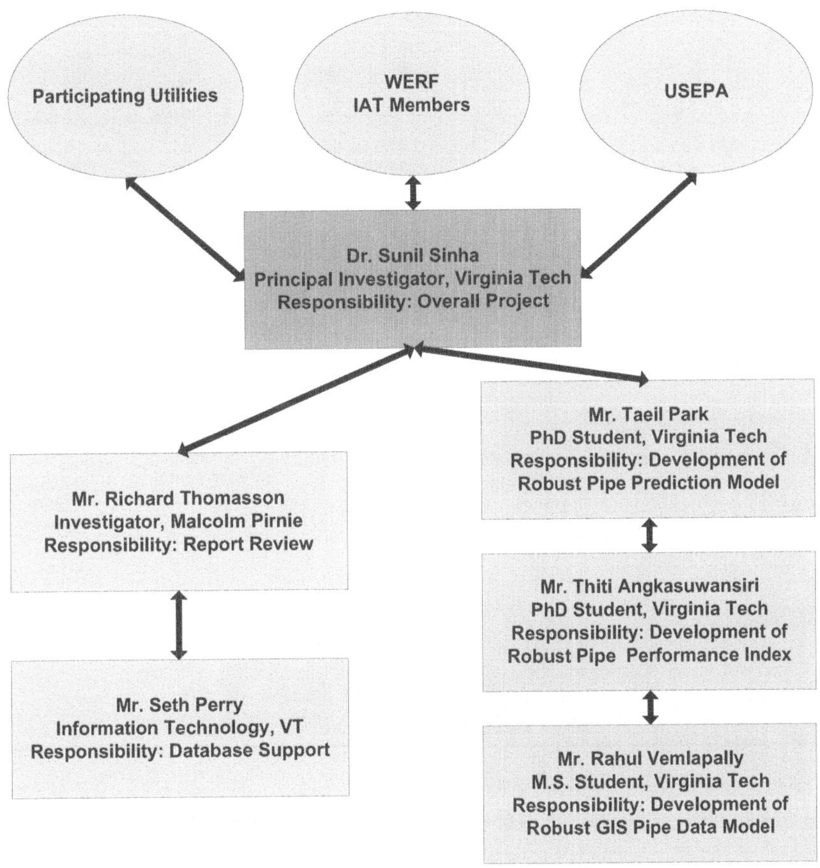

Figure 1-4. Project Organizational Chart.

Participating Utilities

Figure 1-5 shows the utilities across the nation participating in this research project. At present, the following utilities were selected for prototype geographic information system (GIS) database development in the short-term phase of this research project (star marked in figure).

- City of Atlanta, Atlanta, GA
- Seattle Public Utility, Seattle, WA
- Orange County Sanitation District, Fountain Valley, CA
- Pittsburgh Water and Sewer Authority, Pittsburgh, PA
- VPI Sanitation Authority, Blacksburg, VA

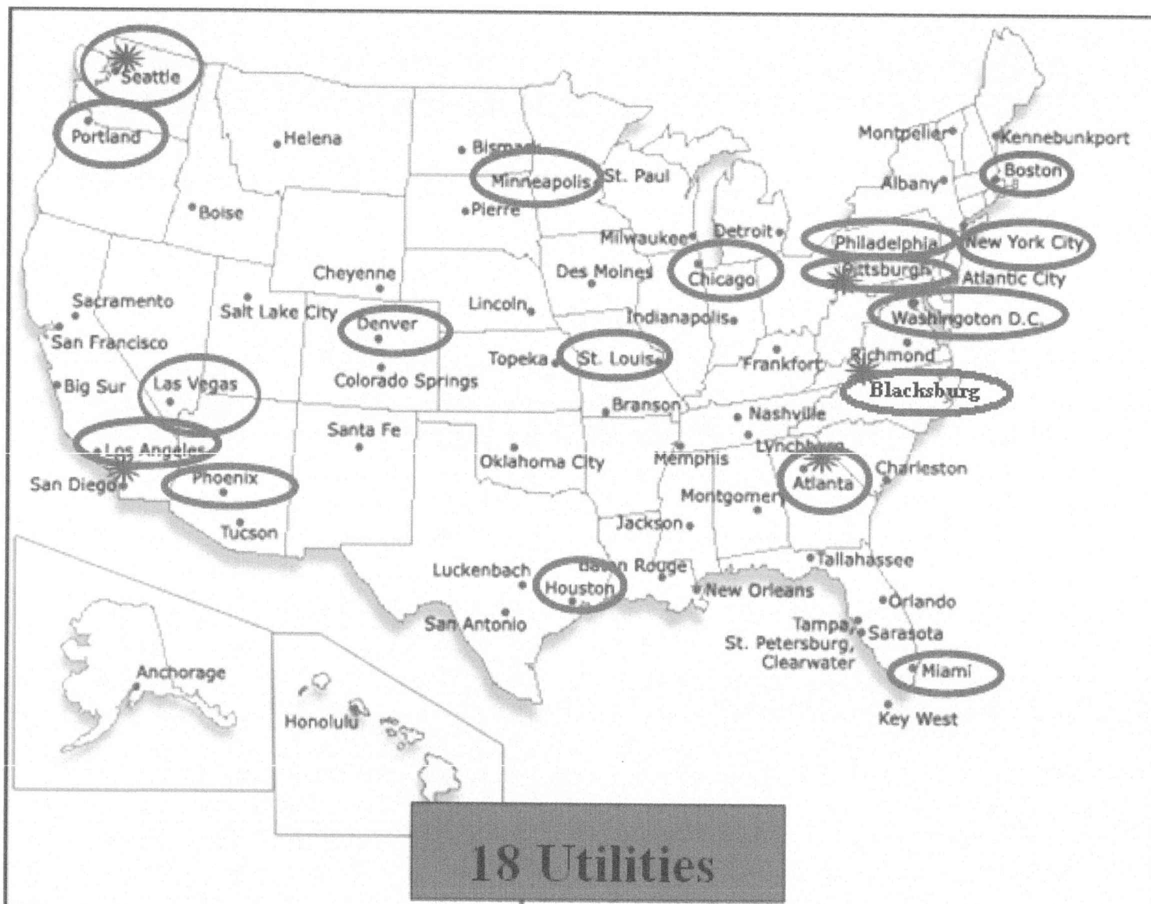

Note: Eighteen utilities have agreed to participate in the research project (circled). At present, five utilities were selected for prototype GIS database development in the short-term phase of this research project (asterisks).

Figure 1-5. Participating Utilities.

CHAPTER 2.0

BACKGROUND

Chapter Highlights

This chapter presents background and current information related to wastewater infrastructure asset management and pipe infrastructure systems.

- **Wastewater Infrastructure System:** Collection and transmission systems, wastewater treatment and sludge management systems, and reuse and recycling systems.
- **Collection and Transmission Systems:** Designed to collect and transport wastewater from domestic, commercial, and industrial sources to treatment plants for processing.
- **Wastewater Treatment:** operations and processes of how wastewater is treated in the plant before being discharged to protect human health and the environment,
- **Sludge Management:** Sludge (also called biosolids) – a semi-solid residual of wastewater treatment, needs additional processing before going to land treatment or sludge landfills.
- **Water Reuse:** Methods to reuse water, which has almost become necessary for conserving and extending available water supplies.
- **Pipe Infrastructure:** Pipelines used in municipal wastewater systems, are made of different types of material depending on suitability of the pipe for location and design purposes.
- **Wastewater Systems in the United States:** More than 19,500 municipal sanitary sewer collection systems nationwide, approximately 500,000 miles of municipally owned pipes in publicly owned systems, and another 500,000 miles of privately owned pipes that transport wastewater into these collection systems.
- **Design and Installation:** Different pipe materials have different design and installation. This section presents an overview of design and installation procedures recommended by pipe associations and research studies.
- **Condition Assessment:** Discusses NASSCO and Water Research Center (WRc) condition assessments of wastewater pipe and other research work.

2.1 Wastewater Infrastructure Asset Management

Drinking water treatment plants use sources such as surface water or groundwater. This water is purified to potable water and then distributed for residential, commercial, and industrial uses. Later, the wastewater from the users is collected and transported to the wastewater treatment plant. Finally, the treated water is sent back to the nearby waterbodies. This municipal water cycle is illustrated in Figure 2-1.

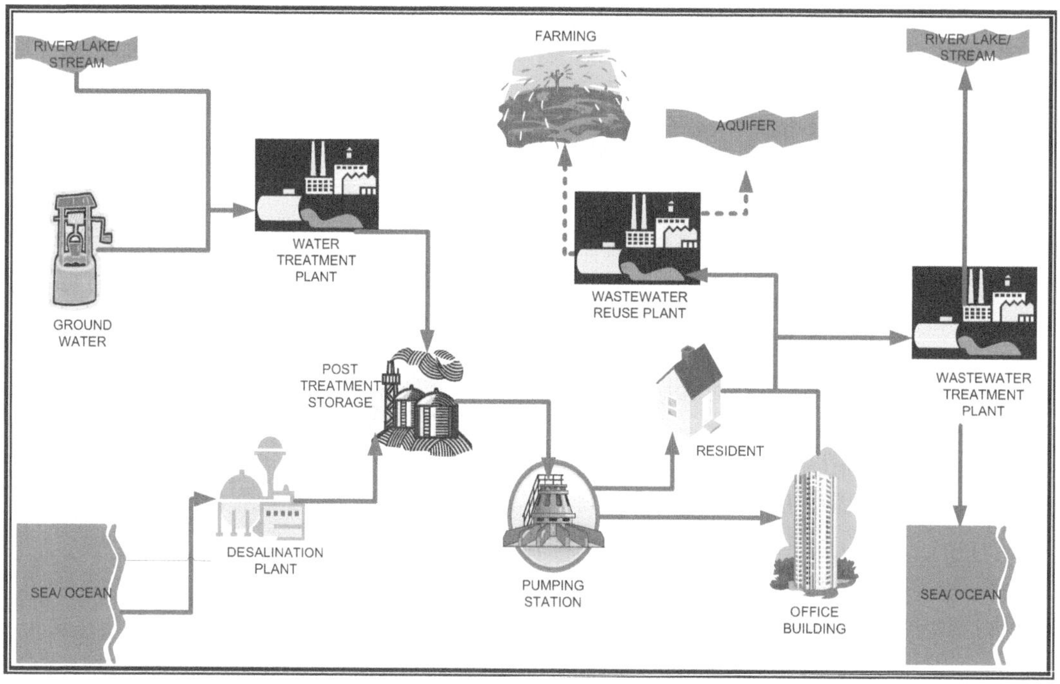

Figure 2-1. Water (Drinking Water and Wastewater) Infrastructure System.

The water infrastructure (drinking water and wastewater) was built to provide the public with access to potable water and sanitation. The wastewater infrastructure system is composed of collection and transmission systems, wastewater treatment and sludge management systems, and reuse/ recycling systems. It collects, transmits, treats, and disposes of water supplies used by domestic, industrial, commercial, and public users [6].

2.1.1 Collection and Transmission Systems

Wastewater collection and transmission systems are designed to collect and transport wastewater from domestic, commercial, and industrial sources to treatment plants for processing [7]. The process begins when service laterals or sewer pipes convey the wastewater away from the buildings where it connects to a sewer main. The sewer main then transports the wastewater to the treatment plant or a pump station where it is pumped into the treatment plant. Once treated, the water is discharged into a stream or a larger body of water. Collection systems in large communities may be made up of pipes that convey the wastewater by means of pumps and gravity. The design of a gravity system is to have an operating velocity of at least 2 ft/s to preclude the deposition of solids. Some of the main considerations for the design of a wastewater pipe are the pipe–soil interaction, slope, materials resistance to scour, and effective joints that minimize or eliminate any infiltration of groundwater into the pipe or exfiltration of the wastewater to the surrounding soils and groundwater areas [8].

Wastewater usually flows under gravity, however pressure flows do exist. Types of pressure flow are:

- Force mains – sewage is pumped where gravity flow is not possible
- Pressure sewers – each household has pump that discharges to the pressure main
- Vacuum sewers – the vacuum pump pulls the flow through the system

The collection system of the wastewater pipelines consists of separated sewers and combined sewers, where combined systems collect stormwater and sanitary.

2.1.2 Wastewater Treatment System

Wastewater is treated in the treatment plant before being discharged to the environment so that the disposed water does not contaminate the environment and harm the public health. In general, the treatment system aims to remove floating material, biodegradable organics, and pathogenic organisms. To address public health concerns, toxic chemicals also have to be suspended. Wastewater treatment plants are classified as physical, chemical, or biological. Typical operations and processes of each unit are listed below [9].

- Physical – screening, mixing, flocculation, sedimentation, flotation, filtration, gas transfer
- Chemical – precipitation, adsorption, disinfection
- Biological process – various biological processes include activated sludge, trickling filter, stabilization ponds

Wastewater treatment removes the following major contaminants: suspended solids, biodegradable organics, volatile organics, pathogens, nutrients, refractory organics, heavy metals, and dissolved organic solids. Industrial wastes may contain additional elements such as acids, high temperatures, toxic chemicals, oils, and greases that must be removed in the pretreatment process before going into wastewater plant. Most wastewater treatment plants in the United States treat less than 1 million gallons per day (mgd), and are considered to be small plants. Smaller wastewater systems, typically 0.01- 0.25 mgd, contain package plants or on-site systems, mainly septic tanks. The most conventional types are extended aeration, contact stabilization, sequencing batch reactors, rotating biological contactor, and physical/chemical. On-site systems are also available for individual homes or cluster of homes [9].

2.1.3 Sludge Management System

Sludge is a semi-solid residual of the wastewater treatment process. It can be categorized into raw primary sludge (fecal material) and secondary sludge (living culture organisms that help remove contaminants from wastewater). Before going to a landfill, sludge needs additional processing, as follows: pumping, grinding, degritting, thickening, stabilization, conditioning, disinfection, dewatering, drying, thermal reduction, and ultimate disposal [9].

2.1.4 Reuse/Recycling System

Water reuse, also called water recycling and water reclamation, is the process by which water is recycled and reused in the natural water cycle. For purposes of this report, it refers to projects that use technology to speed up the natural processes [10]. Water reuse has almost become necessary for conserving and extending available water supplies. In addition, water reuse offers communities an alternate wastewater disposal method and provides pollution abatement by diverting effluent discharge away from sensitive surface waters [11]. As long as it is sufficiently treated, the recycled water can satisfy most water demands.

The U.S. Environmental Protection Agency (U.S. EPA) provides suggested treatment and uses for recycled water, such as crop irrigation, industrial cooling processes, toilet flushing, and car washing. However if human exposure to recycled water is likely, further treatment may be required to prevent disease [10]. This report focuses on the wastewater pipe infrastructure system – the main component of the collection and transmission systems.

2.2 Wastewater Pipe Infrastructure System

Pipelines are the main component in wastewater collection systems. Most buried pipeline networks have been installed using open trench construction methods [12]. This construction method typically consists of placing pipes on bedding material and backfilling the trench. Studies have shown that the structural performance and behavior of the buried pipe depends on the type of backfill placed around the pipe, construction sequence, compaction control, surface loads, and the type of pipe material (flexible or rigid) [13-18].

The pipelines used in municipal wastewater systems are made of different types of material depending on suitability of the pipe location and design purposes. Each pipe material undergoes failure in a different way. Rigid pipe is designed to resist external loads by its inherent strength, whereas flexible pipe relies on the capacity of the surrounding soil to carry the load and provide stability. All types of pipe can perform well, but the conditions for satisfactory long-term performance vary. Furthermore, the performance criteria differ depending on pipe types: the severity of cracking is the main criterion for rigid concrete pipe, whereas the degree of deflection is the main criterion for flexible pipe. For the purpose of analysis, design, or both, it is necessary to develop a complete understanding of the failure modes and mechanisms of buried pipe infrastructure systems.

There are many parameters affecting wastewater pipe infrastructure systems and their failure. Examples of these parameters are structural such as pipe diameter, age, and material, and environmental such as soil properties and external loading. Figure 2-2 shows examples of some of these parameters that affect the pipe deterioration in both short-term and long-term. Changes in one parameter will also affect the others in many ways. For example, in concrete pipe excessive loadings along with poor pipe bedding can cause pipe cracks and fractures. Cracks and fractures in the pipeline will cause infiltration and exfiltration to and from the environment.

2.2.1 Overview of U.S. Wastewater Collection and Transmission Systems

The 1998 Clean Water Needs Survey [19] identified more than 19,500 municipal sewer collection systems nationwide, approximately 500,000 miles of municipally owned pipes in publicly owned systems and another 500,000 miles of privately owned pipes that transport wastewater into these collection systems.

The American Society of Civil Engineers (cooperating with U.S. EPA) [20] conducted a survey of 42 wastewater utilities that indicated the age of collection systems ranged from new to 117 years, with an average age of 33 years (illustrated in Figure 2-3). About 18% of sewers were built in the last 10 years, 41% in the last 20 years, 82% in the last 50 years, and 98% in the last 100 years [21].

By evaluating population, urban density, and public sewerage system data from the U.S. Census Bureau, U.S. EPA [22] estimated the sewer pipe installed per decade. The installation of pipe has followed demographic increase in population and growth in metropolitan areas associated with suburbanization. The census expected an increase in population from 282 million in 2000 to more than 335 million by 2020. The systems will need to keep up with population growth as well as mobilization of people.

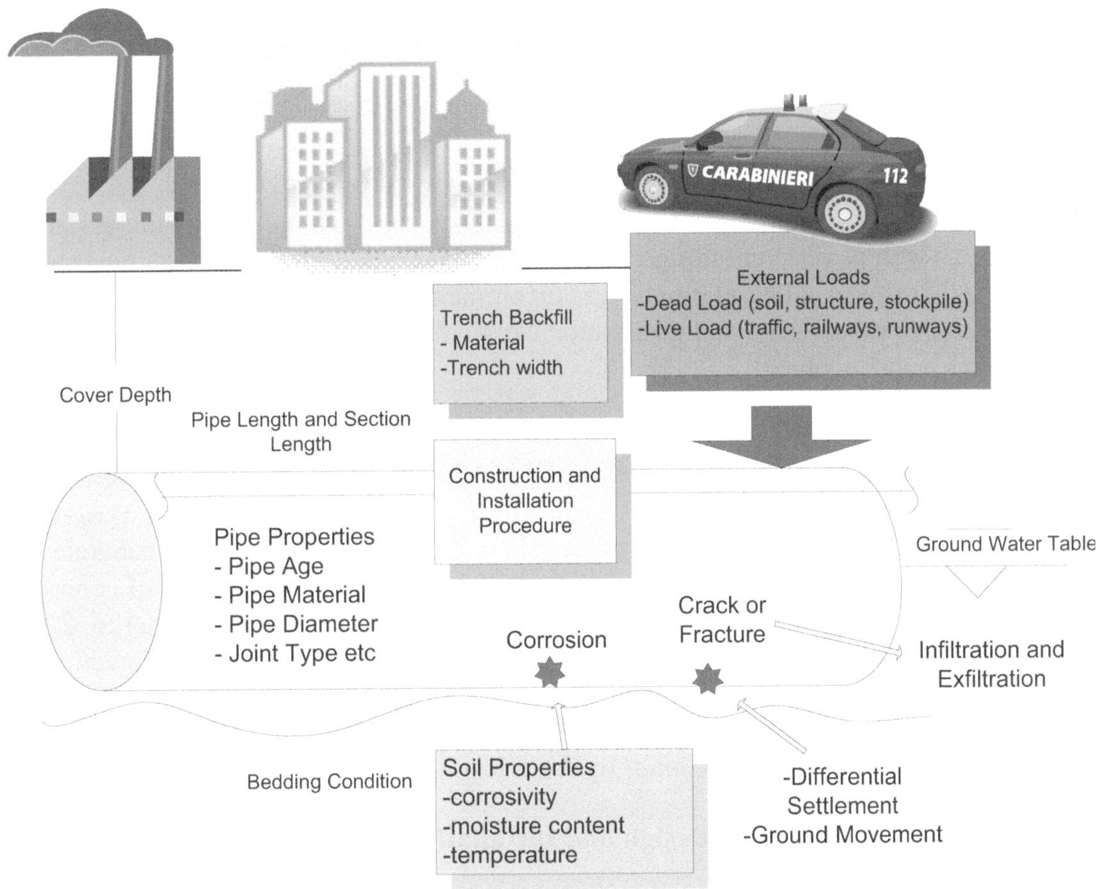

Figure 2-2. Factors Affecting the Condition and Performance of Buried Pipes.

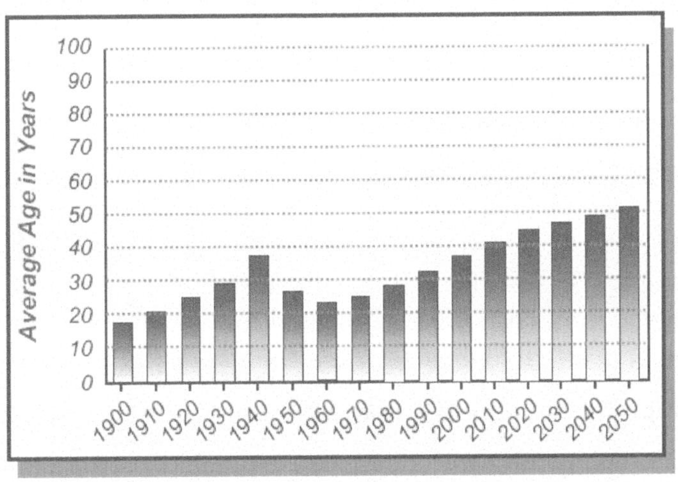

Figure 2-3. Average Age of Wastewater Pipe in Years [22].

Referring to a study performed by the Urban Institute [23], approximately 50 major main breaks and 500 stoppages occur per 1,000 miles per year, amounting to an estimated 35,000 breaks and 350,000 stoppages annually [24]. The problems caused by deterioration of joint materials, pressure surges, disturbance by construction or direct tapping, and seismic activity conclude about 75% of the nation's piping systems functioning at 50% of capacity or less [24, 25].

2.2.2 Wastewater Pipe Design and Installation

Wastewater pipes are made of various materials, each of which has advantages and disadvantages, as well as different design and installation standards. This section gives an overview of design and installation procedures recommended by pipe associations and research studies. The detailed pipe design and installation can be found in handbooks and manuals referenced in each subchapter.

2.2.2.1 Concrete Pipe Design and Installation

Concrete wastewater pipe was developed during 19th century. It is known for its durability characteristics. Non-reinforced concrete pipes make up the majority of most local small diameter (12-18 inches) wastewater pipes. There are two general types of concrete pipe construction; embedded-cylinder and lined-cylinder construction. The benefits of concrete pipe are high beam strength and rigidity, easy joining system, and corrosion resistance [26, 27]. There are three typical types of concrete pipe installation techniques: trench, positive projecting embankment, and negative projecting embankment. In addition, jacking or tunneling method is also applicable where deep excavation is required. Procedures of concrete pipe installation are available in *Concrete Pipe Design Manual* by American Concrete Pipe Association [28].

2.2.2.2 Reinforced Concrete Pipe Design and Installation

Early 19th century, reinforced concrete pipes were developed in the United States. Steel wires are used to increase tensile strength of the pipe. Cage machines, mandrels, and wire rollers are the three common means of fabricating cages [28]. Reinforced concrete pipes are used in gravity lines and diameter ranges from 15-48 inches. Installation procedures are similar to concrete pipe and can be found in *Concrete Pipe Design Manual* by American Concrete Pipe Association [28].

2.2.2.3 Clay Pipe Design and Installation

Vitrified clay pipes have been present in the pipeline industry for a century. One of the first American clay sewer pipe plants was operated in 1849. Clay pipes are strong, chemical, and corrosion resistant, and also heat resistant [27]. An open trench is a typical installation of clay pipe, but microtunneling is also suitable for clay pipe installation. Microtunneling methods offer the least disruptive and the most cost effective method for clay pipe installation. Installation details of clay pipe can be referred to the *Clay Pipe Handbook* [29].

2.2.2.4 Ductile Iron Pipe Design and Installation

In 1975, ductile iron pipe was introduced to the pipeline construction industry. Ductile iron has high strength, durability, low maintenance, and corrosion resistance [26]. The installation of ductile iron pipe is not complicated. It can be installed in various size trench and bedding conditions. Also it can be simply cut in the field. The diameters of pipe vary from 3-64 inches. Cement mortar is normally used for lining finish. Detailed installation procedures can be referred to in the *Installation Guide for Ductile Iron Pipe* by Ductile Iron Pipe Research Association [30, 31].

2.2.2.5 Cast Pipe Design and Installation

Around 1900, grey cast iron pipes were introduced into the market and were produced by casting molten iron in vertical sand molds. Today, many utilities worldwide still have a very high proportion of this type of pipes. Because of misalignment of the central core molds, many of these pipe walls did not have a uniform thickness.

Spun grey iron pipes where then introduced into the market around 1930 and were widespread till 1960. This process included having molten cast iron poured into cylindrical molds made of metal or sand lined, which were rotated at high speeds so the pipe walls were formed by centrifugal force. Most of the pipes ranged from 3-24 inches in diameter. Despite the fact that cast iron pipe is inexpensive, it is no longer manufactured, having been replaced by ductile iron pipe, which possesses higher strength.

2.2.2.6 Polyvinyl Chloride (PVC) Pipe Design and Installation

Polyvinyl chloride (PVC) pipe was introduced and available to use for sanitary sewer and water distribution markets in 1950s. The advantages of PVC pipe are lightweight, easy handling, corrosion resistance, chemical resistance, and flexibility [26]. PVC pipe can be separated into pressure and non-pressure pipe. The installation procedures of PVC pressure pipe include:

- trench construction
- appurtenances
- locating, leak detection, and thawing procedures
- inspection and testing of pipe system

For PVC non-pressure pipe, the installation processes are:

- trench construction
- pipe laying
- appurtenances
- inspection and testing of pipe system

Details of PVC pipe installation are also available in the *Handbook of PVC Pipe Design and Construction* by Uni-Bell PVC Pipe Association [32].

2.2.2.7 Polyethylene Pipe Design and Installation

Introduced in the late 1950s, polyethylene pipe is suitable for rehabilitation, narrow trenching and trenchless applications. The advantages of using polyethylene pipe are flexibility, lightweight, toughness, and corrosion resistance [26].

Installation techniques for polyethylene pipe vary by location and function of the pipe. For example, underground installation of polyethylene pipe can be done by trench excavation. Special installation techniques include plowing, planting, pulling pipe into a narrow trench, pipe bursting, insertion renewal, horizontal boring, and directional boring, all of which require less excavation. Additional information on polyethylene pipe installation is available in the *Manual of Water Supply Practices M55, PE Pipe – Design and Installation* by American Water Work Association [33].

2.2.2.8 Pre-Stressed Cylinder Pipe Design and Installation

In 1942, pre-stressed concrete cylinder pipes were first developed in the United States. The pipes are constructed with a steel cylinder lined with a concrete core or a steel cylinder embedded within a concrete core that is placed by a centrifugal process, radial compaction, or by vertical casting. Once the core is cured, the pipe is helically wrapped with high strength wire with a stress of 75% of the minimum specified tensile strength. This wire is then entrenched into thick cement slurry and coated with mortar. Pre-stressed concrete cylinder pipes are used for force mains. Installation procedures are available in Concrete Pipe Design Manual by American Concrete Pipe Association [28].

2.2.2.9 Steel Pipe Design and Installation

One of the main concerns of steel pipes is corrosion; therefore, these pipes must be appropriately protected against internal as well as external corrosion. Most cases of internal corrosion can be prevented by the application of bitumen, cement mortar, or concrete. External corrosion can be prevented by bitumen application, sheathing or enamel, coal tar wrapping, or polyethylene cladding. Extra care must be taken to ensure the linings are not damaged during transportation of these pipes. Cathodic protection is also recommended.

2.2.2.10 Brick Sewer Design and Installation

Brick masonry was a commonly used material in the early developments of sewer infrastructure. Most sewers in the early 20th century were combined sewers and the majority of brick masonry constructions are for gravity flow. Brick sewers usually consist of one or two half-brick rings bedded on the excavation that are cut to shape. When large brick sewers are constructed in a trench, they usually consist of a brick lining backed by concrete. Brick work makes an excellent lining, because good-quality drainage bricks laid with small joints resist scour.

2.2.3 Wastewater Pipe Condition Assessment

One of the needs of condition assessment is to establish the current condition of assets as a means for prioritizing maintenance and rehabilitation efforts. Some assets are more important than others and should therefore receive proportionally more attention. Condition assessment can be used to understand the level of asset deterioration and the impact it has on the system. The utility can then attempt either to reduce the probability of failure through some operational or capital intervention, or accept the level of risk associated with the asset's condition [34].

This research work builds on the research projects done by WERF and other organizations. Key research projects related to pipe condition assessment that are background for this work are discussed below.

2.2.3.1 CARE-S: Decision Support Tools for Sustainable Water Network Management

CARE-S aims to establish a framework for sewer network rehabilitation decision making. The European Commission developed CARE-S to produce a decision support system that enables municipal engineers to establish and maintain effective management of their sewer networks.

CARE-S takes into account all aspects of rehabilitation decisions with the link to performance indicators. This tool deals with information relevant to wastewater and stormwater network rehabilitation. The performance indicators were developed based on the information from 32 project partners and end-users. There are 41 performance indicators with a large number of utility information and external information.

Analytical and statistical tools to assess and forecast some of the performance indicators were developed as well. There are structural condition tools that are comprised of network level models and specific-pipe level models for corrosion, load bearing capacity, blockage, infiltration, and exfiltration.

Socio-economic and environmental risks of malfunctioning wastewater systems were taken into account in the study. Finally, a prototype software package that supports user decision making in choosing the right rehabilitation technology planning strategy for rehabilitation was proposed [35].

2.2.3.2 NASSCO's Pipeline Assessment and Certification Program

To evaluate the failures of the wastewater pipeline systems, there is a need for a standard format to record failure or defect. Many utilities in the United States have already adopted NASSCO Pipeline Assessment and Certification Program (PACP), a standardized system to evaluate internal television inspections of wastewater collection assets. Visual defects are identified using key structural and, operation and maintenance parameters. Collection system assessment through PACP, or other similar methods, can then be recorded and evaluated. The condition of the wastewater collection system assets is based upon the rate of deterioration of the installed materials. Deterioration factors include, but are not limited to, soil conditions, groundwater, hydraulics, surface loadings, installation and internal conditions (roots, grease, surcharging, etc.). Pipeline defects tend to become more severe over time; such as deformation, subsidence or hydrogen sulfide attack [36].

Televised pipelines are categorized under PACP using four categories:

1. Structural defects: Structural defects are separated into 12 different groups – crack, fracture, broken, hole, deformed pipe, collapse, joint defects, surface damage, lining defect, weld failure, point repair, and brickwork.
2. Operational/maintenance defects: Operational/maintenance defects are separated into five different groups – deposits, roots, infiltration, obstruction, and vermin.
3. Construction features: Construction features are separated into four different groups – tap, intruding sealing material, line, and access points.
4. Other: The other category includes miscellaneous observations.

Table 2-1 shows example of structural condition distress terms defined by NASSCO [36].

Table 2-1. Sewer Pipe Structural Condition Evaluation.

Pipe condition	Description
Collapsed pipe	Complete loss of structural integrity of the pipe as a result of fracturing and collapse of the pipe walls. Most of the cross- section is lost to flow
Structural cracking with deflection	Pipe wall displacement plus cracks
• Longitudinal	Defect runs approximately along the axis of the sewer
• Circumferential	Defect runs approximately right angles to the axis of the sewers
• Multiple	Combination of both longitudinal and circumferential defects
• Slab-out	A large hole in the sewer wall with pieces missing
• Sag	The pipe line invert drops below the downstream invert
Structural cracking without deflection	Sewer wall cracked longitudinally, circumferentially, or multiply, but not displaced
• Cracked joint	The spigot and/or bell of a pipe is cracked or broken
• Open joints	Adjacent pipes are longitudinally displaced at the joint
• Holes	A piece of a pipe wall or joint is missing
• Root intrusion	Tree or plant roots have entered the sewer through an opening in the pipe wall or joint
• Protruding joint material	Joint sealing material or gasket is displaced into the sewer form its original location
Corrosion	When the cementitious pipe material shows evidence of deterioration illustrated by following stages
• Condition1	The pipe wall surface shows irregular smoothness, i.e., wall aggregate is exposed
• Condition 2	The reinforcing steel is exposed
• Condition 3	The reinforcing steel is gone and/or the pipe wall is no longer intact revealing the surrounding soil
Pulled joint	Adjacent pipe joints are deflected beyond allowable tolerances so that the joint is open
Vertical displacement	The spigot of the pipe has dropped below the normal joint closure

2.2.3.3 Water Research Center

Utilities rely on the use of CCTV for inspection and the WRc coding (Table 2-2) for assessing and reporting current pipeline condition. The data collected and coded during CCTV inspection is used as an input of the WRc internal grading system. This system derives a grading of the sewer based on the observation during the CCTV inspection.

Table 2-2. WRc Coding and Standards.

Grade	Description	Implication	Typical defect description (Concrete and clay pipe)
1	Intact	Acceptable structural condition	No structural defect
2	Light deterioration	Minimal collapse risk in short term but potential for further deterioration	Circumferential crack Moderate joint defect, i.e., open joint (medium) of joint displaced (medium) Spalling slight Wear slight
3	Medium Deterioration	Collapse unlikely in the near future but future deterioration likely	Fracture with deformation <5% Longitudinal cracking or multiple cracking Minor loss of level More severe joint defects Spalling medium Wear medium
4	Severe Deterioration	Collapse likely in foreseeable	Broken Deformation up to 10% and broken Fracture with deformation 5–10% Multiple fracture Serious loss of level Spalling large Wear large
5	Collapse	Collapse or collapse imminent	Already collapsed Deformation > 10 % and broken Extensive areas of fabric missing Fracture with deformation > 10%

2.2.3.4 Condition Assessment Protocols Project (WERF)

This research (WERF Project 03-CTS-20CO) provides a road map for the effective use of condition and performance assessment tools within the context of different asset management sophistication levels. Generic approaches to assessment program design and tool selection are offered, which can be applied across a range of asset types. The report recommends the assessment program development draw on various protocols reviewed in the research and case study examples from leading utilities throughout the world [34].

The condition assessments should ideally be undertaken only when required to fill a specific gap in asset-related information to improve decision making. The recommended approach uses an exclusion process in which tools are excluded on the basis of criteria relating to technical feasibility, technical suitability, and utility technical capacity. Useable options must then be evaluated through economic or financial analysis such that final tool selection is made with regard to available resources, the cost-benefits accrued and utility affordability issues [34].

This WERF research provides descriptions and reviews of 83 individual condition assessment tools and techniques used in the water and wastewater industry, including a discussion of principles, applications, practical considerations, advantages and limitations. A prototype expert system was developed to facilitate the selection of condition assessment tools using software systems. Such an expert system could provide a framework for updating, maintaining and distributing refinements of these tool reviews as condition assessment technologies and understanding advance in the future [34].

2.2.3.5 Assessment and Evaluation of Storm and Wastewater Collection Systems (National Research Council, Canada)

It is difficult to manage and follow the development of storm and wastewater collection system maintenance and renewal programs in Canadian municipalities. This can be attributed partly to the unorganized system and the responsive approach followed for the problems that have occurred. The National Research Council's best practices report [37] presents systematic and proactive approaches for the assessment and evaluation of stormwater and wastewater collection systems.

There are five tasks recommended in this research. First, the detailed inventory of storm- and wastewater collection system attributes should be compliant and up to date. The database and guidelines should include information on location, physical dimensions, related land use areas, operating conditions, and applicable operational data. Second, the system operating conditions should be investigated for consistency and accessibility of accurate data. Third, the condition assessment of pipes in the system should be rated based on structural integrity, functional integrity, and hydraulic adequacy. Fourth, on a system-wide basis the individual pipe should be evaluated for performance in all rating criteria. Fifth, a rehabilitation/replacement plan should be developed following the needs. In addition, monitoring, inspection, and detailed assessment of infrastructure in critical sectors are recommended for sewer rehabilitation and replacement work for renewal programs [37].

2.2.3.6 Decision-Support Tools for Predicting the Performance of Water Distribution and Wastewater Collection Systems (U.S. EPA)

Water and wastewater infrastructure systems represent a major capital investment; utilities must ensure they are getting the highest yield possible on their investment, both in terms of dollars and water quality. More accuracy of the utilities' information means more efficiency of the operation, maintenance, and rehabilitation of the systems. This study reviews the European's research and literature reviews that apply a rehabilitation management model. The research topics that were reviewed include; the capabilities and the data required to run eight pipe assessment software applications or models; the infrastructure performance indicators used by three European water authorities; and an approach to collect the necessary performance indicator data based on assessment of the European experience. The study suggested that pipe material, pipe age, section length, number of breaks or bursts, and diameter are minimum data components needed to develop a failure list of pipes from models because the cost of data collection by using models are high.

The research also presents a number of failures in geographic assemblies. There are four models – AssetMap, Gemini VA, KureCad, and UtilNets – integrated a GIS user interface. The study indicates that using the performance indicators as a management tool is limited in European countries. Furthermore, the study of the additional data collection cost versus the benefits of additional system serviceability in the United Kingdom has not been done yet. For the United States, the performance measurement system from the United Kingdom can be enhanced and provided as a model.

The study also recommends developing a web-based survey of state and local government officials, and academic and professional groups to build up a statistically important database of infrastructure performance indicators [38].

CHAPTER 3.0

Wastewater Pipe Failure Mode and Mechanisms

Chapter Highlights

This chapter presents life cycle of wastewater pipe infrastructure systems and a detail description of pipe failure modes and mechanisms:

- **Defects/Damages:** The wastewater pipe may not have 100% quality when installed in the ground. Some defects and/or damage may have taken place during the design and manufacturing processes that ultimately lowers the quality of the overall product.
- **Life Cycle:** The life cycle of wastewater pipe consists of design, manufacture, construction, operation and maintenance, and repair/rehabilitation/replacement. A complete exhaustive life cycle of the wastewater pipe based on material is presented in this chapter.
- **Repair/Rehabilitation/Renewal:** The proper repair/rehabilitation/renewal of failures depends on knowing the causes and selecting the appropriate procedures that take these causes into account; otherwise, the repair may only be temporary.
- **Failure Modes and Mechanisms:** Failure mode of wastewater pipes can be defined as each type of failure that occurs within the pipe. The failure mechanism is an event that causes the pipe to reach one of combined strength and serviceability limit states. Modes and mechanisms of pipe failure in wastewater infrastructure systems are presented based on material type.

3.1 Description of Pipe Life Cycle

The life cycle of material systems consists of their design, development, construction, operation and maintenance, and repair/rehabilitation/replacement. Technical systems and man-made products such as the infrastructure system and the pipes may be viewed as having a certain life cycle. These systems or products can be considered some sort of living system of organisms since they come to being, they grow and interact with their environment, they age, and eventually they die. All repairs, rehabilitations, and renewals can be considered external influences that will affect the life cycle of the system and products. These living systems or organisms have a common behavior, which is commonly known as the bathtub theory. This bathtub theory is a function of the probability of failure with time. The name "bathtub" comes from the line commonly produced by the probability curve [39]. A representation of the bathtub theory as related to the failure probability of piping systems can be seen in Figure 3-1.

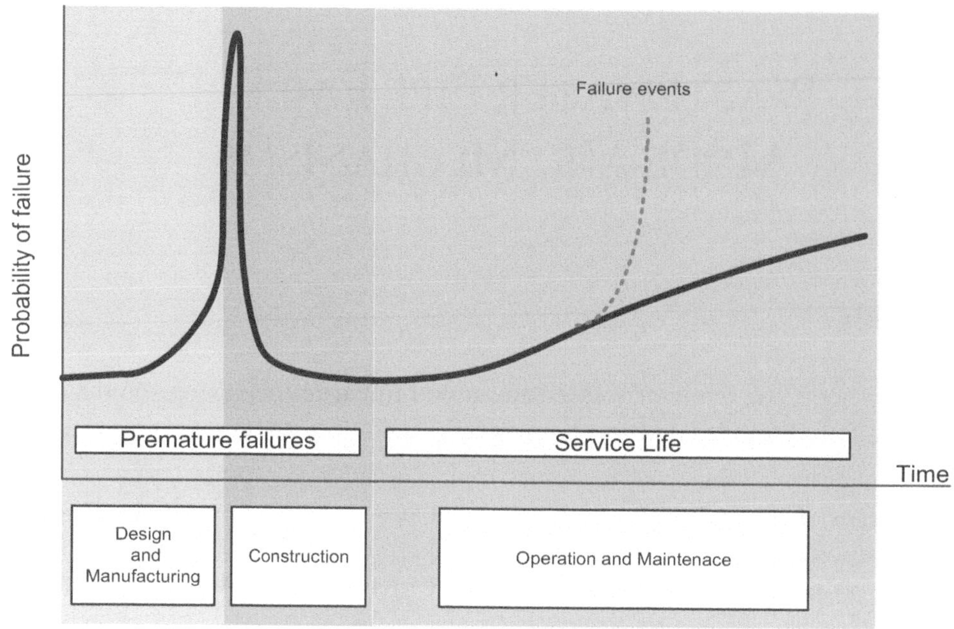

Figure 3-1. Bathtub Theory.

The bathtub theory illustrates that the wastewater pipe may not have 100% quality when installed in the ground. Some defects and damage may have taken place during the design and manufacturing processes, which ultimately lowers the quality of the overall product. Some causes of failure during the design and manufacturing stages are poor design, poor project planning, dimensioning, observation and quality control, manufacturing defects, and storage. The construction process may also have a permanent effect on pipe failure; examples include failure caused by transit, human error, and poor workmanship. Careless or improper construction processes may also lower the performance of pipe. Throughout years of service, operation and maintenance will also affect the pipe performance through various failure causes such as mechanical, thermal, chemical, biological, external interferences, natural catastrophes, inappropriate services, and maintenance.

3.1.1 Life Cycle of Pipe by Material

The research team identified failure causes based on pipe life cycle. Tables 3-1 through 3-7 show wastewater pipe life cycle by material.

Table 3-1. Concrete Wastewater Pipe Life Cycle.

Failure Causes	
Design	
1. Poor Design	Connection, Vertical connections, Mechanical vulnerable material, Chemical vulnerable material, Trust restraint, Corrosion control, Thickness, Special applications
2. Poor Project Planning	Improper connections, Under-design for load, Time limited design, Poor assumptions of environment
3, Dimensioning	
Manufacture	
4. Manufacturing Defects	Wall thickness, Inclusions of unintended structures, Porosity due to air pockets, Longitudinal surface defects, Cheap composite material, Improper connections
5. Storage	
6. Observation and Quality Control	
Construction	
7. Transit	Transit damage
8. Human Error	Third-party damage
9. Poor Workmanship	Improper connections, Uneven bedding, Poor backfilling Material, impact damage, Lining issues, Poor joint
10. Observation and Quality Control	
Operation & Maintenance	
11. Mechanical	Material properties, Hydraulic factors
12. Thermal	Temperature failures
13. Chemical	Chemical attacks
14. Biological	Biological attacks
15. External Interference	Movement of soil, Tensile and compression failures
16. Internal Interference	Fatigue
17. Natural Catastrophes	
18. Inappropriate Service	
19. Inappropriate Maintenance	
Repair/Rehabilitation/Replacement	
20. Design Error	
21. Manufacturing Error	
22. Construction Error	
23. Operation and Maintenance Error	

Table 3-2.Clay Wastewater Pipe Life Cycle.

Failure Causes	
Design	
1. Poor Design	Thickness
2. Poor Project Planning	Under-design for load, Poor assumptions of environment
3. Dimensioning	
Manufacture	
4. Manufacturing Defects	Porosity caused by air pockets
5. Storage	
6. Observation and Quality Control	
Construction	
7. Transit	Transit damage
8. Human Error	Third-party damage
9. Poor Workmanship	Bell holes, Uneven bedding, Poor backfilling material, Impact damage
10. Observation and Quality Control	
Operation & Maintenance	
11. External Interference	Movement of soil
12. Differential Settlement	
13. Natural Catastrophes	
14. Inappropriate Service	
15. Inappropriate Maintenance	
Repair/Rehabilitation/Replacement	
16. Design Error	
17. Manufacturing Error	
18. Construction Error	
19. Operation & Maintenance Error	

Table 3-3. Ductile Iron Wastewater Pipe Life Cycle.

Failure Causes	
Design	
1. Poor Design	Coating, connection, Vertical connections, Chemical vulnerable material, Thrust restraint, Corrosion control, Thickness, Special applications
2. Poor Project Planning	Improper connections, Under-design for load, Time limited design, Poor assumptions of environment
3. Dimensioning	
Manufacture	
4. Manufacturing Defects	Wall thickness, Inclusions of unintended structures, Surface defects, Cheap composite material, Improper connections
5. Storage	
6. Observation and Quality Control	
Construction	
7. Transit	Coating damage
8. Human Error	Third-party damage
9. Poor Workmanship	Improper connections, Uneven bedding, Poor backfilling material, Impact damage, Lining issues
10. Observation and Quality Control	
Operation & Maintenance	
11. Mechanical	Material properties, Hydraulic factors
12. Thermal	Temperature failures
13. Chemical	Pitting corrosion, Galvanic corrosion, Graphitization
14. Biological	Microbiologically influenced corrosion
15. External Interference	Dissimilar electrolytes, Stray current corrosion, Movement of soil, Tensile and compression failures
16. Internal Interference	Fatigue
17. Natural Catastrophes	
18. Inappropriate Service	
19. Inappropriate Maintenance	
Repair/Rehabilitation/Replacement	
20. Design Error	
21. Manufacturing Error	
22. Construction Error	
23. Operation and Maintenance Error	

Table 3-4. Grey Cast Iron Wastewater Pipe Life Cycle.

Failure Causes	
Design	
1. Poor Design	Coating, Connection, Vertical connections, Mechanical vulnerable material, Chemical vulnerable material, Trust restraint, Corrosion control, Thickness, Special applications
2. Poor Project Planning	Improper connections, Under-design for load, Time limited design, Poor assumptions of environment
3. Dimensioning	
Manufacture	
4. Manufacturing Defects	Wall thickness, Inclusions of unintended structures, Porosity due to air pockets, Longitudinal surface defects, Cheap composite material, Pipe material content, Improper connections
5. Storage	
6. Observation and Quality Control	
Construction	
7. Transit	Coating damage
8. Human Error	Third-party damage
9. Poor Workmanship	Improper connections, Uneven bedding, Poor backfilling material, Impact damage, Lining issues
10. Observation and Quality Control	
Operation & Maintenance	
11. Mechanical	Material properties, Hydraulic factors
12. Thermal	Temperature failures
13. Chemical	Pitting corrosion, Galvanic corrosion, Graphitization
14. Biological	Microbiologically influenced corrosion
15. External Interference	Dissimilar electrolytes, Stray current corrosion, Movement of soil, Tensile and compression failures
16. Internal Interference	Fatigue
17. Natural Catastrophes	
18. Inappropriate Service	
19. Inappropriate Maintenance	
Repair/Rehabilitation/Replacement	
20. Design Error	
21. Manufacturing Error	
22. Construction Error	
23. Operation and Maintenance Error	

Table 3-5. PVC Wastewater Pipe Life Cycle.

Failure Causes	
Design	
1. Poor Design	Connection, Vertical connections, Trust restraint, Thickness, Special applications
2. Poor Project Planning	Improper connections, Under-design for load, Time limited design, Poor assumptions of environment
3. Dimensioning	
Manufacture	
4. Manufacturing Defects	Wall thickness, Inclusions of unintended structures, Porosity due to air pockets, Surface defects, Cheap composite material, Improper connections
5. Storage	
6. Observation and Quality Control	
Construction	
7. Transit	Transit damage
8. Human Error	Third-party damage
9. Poor Workmanship	Improper connections, Uneven bedding, Poor backfilling material, Impact damage, Improper trench
10. Observation and Quality Control	
Operation & Maintenance	
11. Mechanical	Material properties, Hydraulic factors
12. Thermal	Temperature failures
13. Chemical	Chemical attacks
14. Biological	Biological attacks
15. External Interference	Movement of soil, Tensile and compression failures
16. Internal Interference	Fatigue
17. Natural Catastrophes	
18. Inappropriate Service	
19. Inappropriate Maintenance	
Repair/Rehabilitation/Replacement	
20. Design Error	
21. Manufacturing Error	
22. Construction Error	
23. Operation and Maintenance Error	

Table 3-6. Polyethylene and High-density Polyethylene Wastewater Pipe Life Cycle.

Failure Causes	
Design	
1. Poor Design	Connection, Vertical connections, Thrust restraint, Thickness, Special applications
2. Poor Project Planning	Improper connections, Under-design for load, Time limited design, Poor assumptions of environment
3. Dimensioning	
Manufacture	
4. Manufacturing Defects	Wall thickness, Inclusions of unintended structures, Porosity due to air pockets, Surface defects, Cheap composite material, Improper connections
5. Storage	
6. Observation and Quality Control	
Construction	
7. Transit	Transit damage
8. Human Error	Third-party damage
9. Poor Workmanship	Improper connections, Uneven bedding, Poor backfilling material, Impact damage, Improper trench
10. Observation and Quality Control	
Operation & Maintenance	
11. Mechanical	Material properties
12. Thermal	Temperature failures
13. Chemical	Chemical attacks
14. Biological	Biological attacks
15. External Interference	Movement of soil, Tensile and compression failures
16. Internal Interference	Fatigue
17. Natural Catastrophes	
18. Inappropriate Service	
19. Inappropriate Maintenance	
Repair/Rehabilitation/Replacement	
20. Design Error	
22. Manufacturing Error	
23. Construction Error	
24. Operation and Maintenance Error	

Table 3-7. Brick Sewers Life Cycle.

Failure Causes	
Design	
1. Poor Design	Mechanical vulnerable material, Chemical vulnerable material, Wall thickness
2. Poor Project Planning	Under-design for load, Time limited design, Poor assumptions of environment
3. Dimensioning	
Construction	
4. Human Error	Third-party damage
5. Poor Workmanship	Uneven bedding, Impact damage
6. Observation and Quality Control	
Operation & Maintenance	
7. Thermal	Temperature failures
8. Chemical	Chemical attacks
9. Biological	Biological attacks
10. External Interference	Movement of soil, Tensile and compression failures
11. Natural Catastrophes	
12. Inappropriate Service	
13. Inappropriate Maintenance	
Repair/Rehabilitation/Replacement	
14. Design Error	
15. Manufacturing Error	
16. Construction Error	
17. Operation and Maintenance Error	

A more complete description of wastewater pipe life cycle based on material and brief explanations can be seen in Appendix A. A list of attachments for wastewater pipe life cycle is shown in Table 3-8.

Table 3-8. List of Appendices for Wastewater Pipe Life Cycle [8, 13, 15, 36, 40-51].

Appendix	Topic
A-1	Life Cycle of Wastewater Concrete Pipes
A-2	Life Cycle of Wastewater Clay Pipe
A-3	Life Cycle of Wastewater Ductile Iron (DI) Pipe
A-4	Life Cycle of Wastewater Cast Iron (CI) Pipe
A-5	Life Cycle of Wastewater Polyvinyl Chloride (PVC) Pipe
A-6	Life Cycle of Wastewater Polyethylene (PE) and High-density Polyethylene (HPDE) Pipes
A-7	Life Cycle of Brick Sewer

3.2 Description of Failure Modes and Mechanisms

The failure processes in buried pipes are complex. At the basic level, pipe failures are caused by applied forces exceeding the residual strength of the pipe material. In general, the forces applied to buried pipe can be considered as five groups: those produced by internal pressure, bending forces, crushing forces, soil movement-induced tensile forces, and temperature-induced expansive forces. The proper repair of failures depends on knowing the causes and selecting the appropriate repair procedures; otherwise, the repair may only be temporary. The study in this phase examined the durability and performance of buried pipes to help municipal and utility engineers to understand the effects of various factors as well as to determine the best pipe material for specific site conditions.

Failure modes of wastewater pipes can be defined as each type of failure that occurs within the pipe. The failure mechanism is an event that causes the pipe to reach one of combined strength and serviceability limit states [39]. There are two limit states: ultimate limit state and serviceability limit state. The ultimate limit state defines a condition at which the strength of the pipe is reached. Examples of this state may be by loss of water tightness, burst, and loss of stiffness. The serviceability limit state defines a condition at which a particular function of the pipe is no longer fulfilled. Examples of this state may be large deformations, change of color, buckling, clogging, abrasion, and local damages [39].

The research team prepared various modes and mechanisms of pipe failure in wastewater infrastructure system based on extensive literature reviews and pipe association interviews.

The pipe failure modes and mechanisms described in this report are separated into material type due to differing modes and mechanisms. For example ductile iron pipes may fail because of corrosion; however, corrosion is not present in PVC pipes. The failure modes and mechanisms are then classified into two groups; main pipe and joint. All failure modes are listed and are separated into a more specific failure type. For example, the failure mode "crack" can be separated into longitudinal, circumferential, spiral, and multiple based on the material type.

Tables 3-9 to 3-20 present main line and joint wastewater failure modes.

Table 3-9. Failure Modes of Concrete Wastewater Pipe.

Concrete Wastewater Pipe Failure Modes		
		A. Main Line
Typical Failure Mode		
Structural Failure		
1	Crack	Longitudinal crack, Circumferential crack, Multiple cracks
2	Fracture	Longitudinal fracture, Circumferential fracture, Multiple fractures
3	Broken	
4	Hole	
5	Deformed	Vertical deformed, Horizontal deformed
6	Collapsed	
7	Surface Damage	Roughness increased, Aggregate visible, Aggregate projecting, Aggregate missing, Missing wall, Surface spalling
8	Lining Failure	Defective Weld, Detached, Defective end, Blistered, Service cut shifted, Abandoned connection, Overcut service, Undercut service, Buckled, Wrinkled
Operational and Maintenance Failure		
9	Deposits	Attached, Settled, Ingress
10	Roots	Fine, Tap, Medium, Ball
11	Infiltration	Weeper, Dripper, Runner, Gusher
12	Exfiltration	
13	Obstacles/Obstructions	Pipe material in invert, Object intruding through wall, Object wedged in the joint, Object through connection, External pipe or cable, Construction debris, Rocks, Other obstacles
Rare Failure Mode		
Structural Failure		
14	Crack	Spiral crack
15	Fracture	Spiral fracture
		B. Joint
Typical Failure Mode		
1	Lined Joint	Separated joint
2	Rubber Gasket Joint	Corrosion degradation of reinforcing wires , Gasket degradation, Concrete degradation, Mortar seal degradation, Seal displacement
3	Steel Endring Joint	Weld failure, Corrosion of spigot , Ovality of pipe

Table 3-10. Failure Modes of Reinforced Concrete Wastewater Pipe.

Reinforced Concrete Wastewater Pipe Failure Modes		
		A. Main Line
Typical Failure Mode		
Structural Failure		
1	Crack	Longitudinal crack, Circumferential crack, Multiple cracks
2	Fracture	Longitudinal fracture, Circumferential fracture, Multiple fractures
3	Broken	
4	Hole	
5	Deformed	Vertical deformed, Horizontal deformed
6	Collapsed	
7	Surface Damage	Roughness increased, Aggregate visible, Aggregate projecting, Aggregate missing, Reinforcement visible, Reinforcement projecting, Reinforcement corroded, Missing wall, Surface spalling
8	Lining Failure	Defective Weld, Detached, Defective end, Blistered, Service cut shifted, Abandoned connection, Overcut service, Undercut service, Buckled, Wrinkled
Operational and Maintenance Failure		
9	Deposits	Attached, Settled, Ingress
10	Roots	Fine, Tap, Medium, Ball
11	Infiltration	Weeper, Dripper, Runner, Gusher
12	Exfiltration	
13	Obstacles/Obstructions	Pipe material in invert, Object intruding through wall, Object wedged in the joint, Object through connection, External pipe or cable, Construction debris, Rocks, Other obstacles
Rare Failure Mode		
Structural Failure		
14	Crack	Spiral crack
15	Fracture	Spiral fracture
		B. Joint
Typical Failure Mode		
1	Lined Joint	Separated Joint
2	Rubber Gasket Joint	Corrosion degradation of reinforcing wires , Gasket degradation, Concrete degradation, Mortar seal degradation, Seal displacement
3	Steel Endring Joint	Weld failure, Corrosion of spigot, Ovality of pipe

Table 3-11. Failure Modes of Clay Wastewater Pipe.

Clay Wastewater Pipe Failure Modes		
	A. Main Line	
Typical Failure Mode		
Structural Failure		
1	Cracks / Fractures	Longitudinal (localized) crushing failure, Star breaks, Shear failure, Lateral shear, Shear between manhole pipeline
2	Collapsed	
Operational and Maintenance Failure		
3	Sags in the Line	
4	Deposits	Attached, Settled, Ingress
5	Roots	Fine, Tap, Medium, Ball
6	Infiltration	Weeper, Dripper, Runner, Gusher
7	Exfiltration	
Rare Failure Mode		
Structural Failure		
8	Surface Damage	
9	Abrasion	
Operational and Maintenance Failure		
10	Obstacles/Obstructions	Pipe material in invert, Object intruding through wall, Object wedged in the joint, Object through connection, External pipe or cable, Construction debris, Rocks, Other obstacles
	B. Joint	
Failure Mode		
Typical Failure Mode		
1	Parallel Offsets	
2	Pulled Joints	
3	Excessive angular deflection	
4	Joint Leakage	

Table 3-12. Failure Modes of Ductile Iron Wastewater Pipe (Gravity).

Ductile Iron Wastewater Pipe Failure Modes (Gravity)		
		A. Main Line
Typical Failure Mode		
Structural Failure		
1	Hole	
2	Corrosion	External corrosion, Internal corrosion
3	Deformed	Vertical deformed, Horizontal deformed
4	Lining Failure	Detached, Defective end, Blistered, service cur shifted, Abandoned connection, Overcut service, Undercut service, Buckled, Wrinkled, Spall, Crack
Operational and Maintenance Failure		
5	Deposits	Attached, Settled, Ingress
6	Roots	Fine, Tap, Medium, Ball
7	Infiltration	Weeper, Dripper, Runner, Gusher
8	Exfiltration	
Rare Failure Mode		
Structural Failure		
9	Crack	Circumferential crack, Longitudinal crack, Spiral crack, Mixed crack
10	Fracture	Circumferential fracture, Longitudinal fracture, Spiral fracture, Mixed fracture, Shearing, Bell shearing
11	Collapsed	
Operational and Maintenance Failure		
12	Obstacles/Obstructions	Object intruding through wall, Object wedged in the joint, Object through connection, External pipe or cable, Other obstacles
		B. Joint
Typical Failure Modes		
1	Push-On	Gasket/Seal failure, Angular deflection, Axial pull out, Thermal expansion/cracked spigot, Lateral pipe movement, Internal corrosion, External corrosion
2	Bolted Flange	Bolt failure, Flange failure, Gasket failure, Stress at flange , Bolt corrosion
3	Restrained Joint	Pull out, External corrosion
Rare Failure Modes		
4	Mechanical Joint	Bolt failure, Bolt corrosion, Backing ring corrosion, Structural spigot failure
5	Slip-On Coupler	Internal corrosion, External corrosion, Bolt corrosion, Spigot pull-out
6	Ball & Socket	Structural socket failure

Table 3-13. Failure Modes of Cast Iron Wastewater Pipe (Gravity).

Cast Iron Wastewater Pipe Failure Modes (Gravity)	
	A. Main Line
Typical Failure Mode	
Structural Failure	
1. Crack	Circumferential crack, Longitudinal crack, Spiral crack, Mixed crack
2. Fracture	Circumferential fracture, longitudinal fracture, Spiral fracture, Mixed fracture
3. Hole	Blowout holes
4. Corrosion	External corrosion, Internal corrosion
5. Lining Failure	Detached, Defective end, Blistered, Service cut shifted, Abandoned connection, Overcut service, Undercut service, Buckled, Wrinkled, Spall, Crack
Operational and Maintenance Failure	
6. Deposits	Attached, Settled, Ingress
7. Roots	Fine, Tap, Medium, Ball
8. Infiltration	Weeper, Dripper, Runner, Gusher
9. Exfiltration	
Rare Failure Mode	
Structural Failure	
10. Shearing	Bell Shearing
11. Burst Failure	Burst failure (int.), Burst Failure (ext.), Burst Failure (3rd party)
12. Collapsed	
Operational and Maintenance Failure	
13. Obstacles/Obstructions	Object intruding through wall, Object wedged in the joint, Object through connection, External pipe or cable, Other obstacles
	B. Joint
1. Mechanical Joint	Bolt failure, Bolt corrosion, Backing ring corrosion, Structural spigot failure
2. Bolted Flange	Bolt failure, Flange failure, Gasket failure, Stress at flange, Bolt corrosion
3. Slip-On Coupler	Internal corrosion, External corrosion, Bolt corrosion, Spigot pull-out
4. Groove & Shoulder	Structural socket failure
5. Ball & Socket	Structural socket failure
Rare Failure Modes	
6. Push-On	Gasket/seal failure, Angular deflection, Axial pull out, Thermal expansion/cracked spigot, Lateral pipe movement, Internal corrosion, External corrosion
7. Ball & Spigot Lead/Jute	Angular deflection, Axial pull out, Lead/jute deterioration, Thermal expansion/cracked spigot, Lateral pipe movement, External corrosion

Table 3-14. Failure Modes of PVC Wastewater Pipe (Gravity).

PVC Wastewater Pipe Failure Modes (Gravity)		
		A. Main Line
Typical Failure Mode		
Structural Failure		
1	Crack	Longitudinal crack, Spiral crack, Bell splitting
2	Fracture	Longitudinal fracture, Spiral fracture
3	Broken	
4	Hole	
5	Deformed	Axial Deformed
6	Buckling	Axial buckling, Transverse/Ring buckling, Non-symmetric buckling, Longitudinal buckling, Other
7	Color Change	
8	Blisters/Voids	Voids
Operational and Maintenance Failure		
9	Deposits	Attached, Settled, Ingress
10	Roots	Fine, Tap, Medium, Ball
11	Infiltration	Weeper, Dripper, Runner, Gusher
12	Exfiltration	
Rare Failure Mode		
Structural Failure		
13	Crack	Circumferential crack , Mixed crack, Irregular crack, Rapid Crack Propagation (RCP)
14	Fracture	Circumferential fracture, Mixed fractures
15	Crazing	Through crazing, External crazing, Internal crazing
16	Deformed	Ring Deformed
17	Dimension Change	
18	Blisters/Voids	External blister, Internal blister, Blazes, Delamination
19	Abrasion/Erosion	
20	Chemical Attacks	
21	Collapse	
Operational and Maintenance Failure		
23	Obstacles/Obstructions	Object intruding through wall, Object wedged in the joint, Object through connection, External pipe or cable, Other obstacles
		B. Joint
Typical Failure Mode		
1	Push-fit and Slip-on Collar (Gasket)	Displaced gasket (into socket), Cracking of socket
Rare Failure Modes		
2	Push-fit and Slip-on Collar (Gasket)	Spigot deflection, Gasket degradation, Bell splitting, PVC material degradation
3	Solvent Weld	Poor Adhesion degradation, Embrittlement, Bell splitting, Angular deflection, Misalignment
4	Other	Joint leak, Mechanical joint leak, Blown rubber on flange, Blown joint, Leak on collar, Leaking rubber ring, Gasket/Seal failure, Structural bell failure, Butt fusion joint, Electro-fusion joint, Flange joint failure

Table 3-15. Failure Modes of Polyethylene and High-Density Polyethylene Wastewater Pipes (Gravity).

Polyethylene and High-Density Polyethylene Wastewater Pipes Failure (Gravity)	
	A. Main Line
Failure Mode	
Typical Failure Mode	
Structural Failure	
1. Crack	Longitudinal crack, Spiral crack
2. Fracture	Longitudinal fracture, Spiral fracture
3. Broken	
4. Hole	
5. Deformed	Axial
Operational and Maintenance Failure	
6. Deposits	Attached, Settled, Ingress
7. Roots	Fine, Tap, Medium, Ball
8. Infiltration	Weeper, Dripper, Runner, Gusher
9. Exfiltration	
Rare Failure Mode	
Structural Failure	
10. Crack	Circumferential crack, Mixed crack, Irregular crack
11. Fracture	Circumferential fracture, Mixed fractures
12. Deformed	Ring
13. Dimension Change	
14. Abrasion/Erosion	
15. Chemical Attacks	
16. Collapse	
Operational and Maintenance Failure	
17. Obstacles/Obstructions	Object intruding through wall, Object wedged in the joint, Object through connection, External pipe or cable, Other Obstacles
	B. Joint
Typical Failure Mode	
1. Butt Fusion	Poor joint integrity, Gaps/cracks/splits/voids/dirt at weld interface, Mixed SDR materials, Lateral offset or misalignment
2. Electrofusion	Pull out in fitting, Poor joint integrity, Voids/contamination, Overheating of fusion wire
3. Compression Fitting	High local stress, Pull-out, Crack, Insert size - leakage

Table 3-16. Failure Modes of Pre-Stressed Concrete Cylinder Wastewater Pipe.

Pre-Stressed Concrete Cylinder Wastewater Pipe Failure Modes		
		A. Main Line
Typical Failure Mode		
Structural Failure		
1	Crack	Longitudinal crack, Circumferential crack, Multiple cracks
2	Fracture	Longitudinal fracture, Circumferential fracture, Multiple fractures
3	Broken	
4	Hole	
5	Deformed	Vertical deformed, Horizontal deformed
6	Collapsed	
7	Surface damage	Roughness increased, Aggregate visible, Aggregate projecting, Aggregate missing, Reinforcement visible, Reinforcement projecting, Reinforcement corroded, Missing wall, Surface spalling
8	Wire Break	
9	Lining Failure	Defective weld, Detached, Defective end, Blistered, Service cut shifted, Abandoned connection, Buckled, Wrinkled
Operational and Maintenance Failure		
10	Exfiltration	
11	Obstacles/Obstructions	Object intruding through wall, Object wedged in the joint, Object through connection, External pipe or cable, Other obstacles
Rare Failure Mode		
Structural Failure		
12	Crack	Spiral crack
13	Fracture	Spiral fracture
		B. Joint
Typical Failure Mode		
1	Lined Joint	Separated Joint
2	Rubber Gasket Joint	Corrosion degradation of reinforcing wires , Gasket degradation, Concrete degradation, Mortar seal degradation, Seal displacement
3	Steel Endring Joint	Weld failure, Corrosion of spigot , Ovality of pipe

Table 3-17. Failure Modes of Ductile Iron Wastewater Pipe (Force Main).

Ductile Iron Wastewater Pipe Failure Modes (Force Main)		
		A. Main Line
Typical Failure Mode		
Structural Failure		
1	Hole	Blowout holes
2	Corrosion	External corrosion, Internal corrosion
3	Deformed	Vertical deformed, Horizontal deformed
4	Lining Failure	Detached, Defective end, Blistered, Service cut shifted, Abandoned connection, Buckled, Wrinkled, Spall, Crack
Operational and Maintenance Failure		
5	Exfiltration	
Rare Failure Mode		
Structural Failure		
6	Crack	Circumferential crack, Longitudinal crack, Spiral crack, Mixed crack
7	Fracture	Circumferential fracture, Longitudinal fracture, Spiral fracture, Mixed fracture
8	Shearing	Bell shearing
9	Burst Failure	Burst failure (Int.), Burst failure (Ext.), Burst failure (3rd party)
10	Collapsed	
Operational and Maintenance Failure		
11	Obstacles/Obstructions	Object intruding Through wall, Object wedged in the joint, Object through connection, External pipe or cable, Other obstacles
		B. Joint
Typical Failure Modes		
1	Push-On	Gasket/Seal failure, Angular deflection, Axial pull out, Thermal expansion/cracked spigot, Lateral pipe movement, Internal corrosion, External corrosion
2	Bolted Flange	Bolt failure, Flange failure, Gasket failure, Stress at flange , Bolt corrosion
3	Restrained Joint	Pull out, External corrosion
Rare Failure Modes		
4	Mechanical Joint	Bolt failure, Bolt corrosion, Backing ring corrosion, Structural spigot failure
5	Slip-On Coupler	Internal corrosion, External corrosion, Bolt corrosion, Spigot pull-out
6	Ball & Socket	Structural socket failure

Table 3-18. Failure Modes of Cast Iron Wastewater Pipe (Force Main).

Cast Iron Wastewater Pipe Failure Modes (Force Main)	
	A. Main Line
Typical Failure Mode	
1. Crack	Circumferential crack, Longitudinal crack, Spiral crack, Mixed crack
2. Fracture	Circumferential fracture, Longitudinal fracture, Spiral fracture, Mixed fracture
3. Hole	Blowout holes
4. Corrosion	External corrosion, Internal corrosion
5. Lining Failure	Detached, Defective end, Blistered, Service cut shifted, Abandoned connection, Buckled, Wrinkled, Spall, Crack
Operational and Maintenance Failure	
6. Exfiltration	
Rare Failure Mode	
Structural Failure	
7. Shearing	Bell shearing
8. Burst Failure	Burst failure (int.), Burst Failure (ext.), Burst failure (3rd party)
9. Collapsed	
Operational and Maintenance Failure	
10. Obstacles/Obstructions	Object intruding Through wall, Object wedged in the joint, Object through connection, External pipe or cable, Other obstacles
	B. Joint
Typical Failure Modes	
1. Mechanical Joint	Bolt failure, Bolt corrosion, Backing ring corrosion, Structural spigot failure
2. Bolted Flange	Bolt failure, Flange failure, Gasket failure, Stress at flange, Bolt corrosion
3. Slip-On Coupler	Internal corrosion, External corrosion, Bolt corrosion, Spigot pull-out
4. Groove & Shoulder	Structural socket failure
5. Ball & Socket	Structural socket failure
Rare Failure Modes	
6. Push-On	Gasket/seal failure, Angular deflection, Axial pull out, Thermal expansion/cracked spigot, Lateral pipe movement, Internal corrosion, External corrosion
7. Ball & Spigot Lead/Jute	Angular deflection, Axial pull out, Lead/jute deterioration, Thermal expansion/cracked spigot, Lateral pipe movement, External corrosion

Table 3-19. Failure Modes of PVC Wastewater Pipe (Force Main).

PVC Wastewater Pipe Failure Modes (Force Main)		
		A. Main Line
Typical Failure Mode		
Structural Failure		
1	Crack	Longitudinal crack, Spiral crack, Bell splitting
2	Fracture	Longitudinal fracture, Spiral fracture
3	Broken	
4	Hole	
5	Deformed	Axial
6	Leakage/Burst	Burst, Burst main, Burst section, Broken section, Blown section
7	Color Change	
8	Blisters/Voids	Voids
Operational and Maintenance Failure		
9	Exfiltration	
Rare Failure Mode		
Structural Failure		
10	Crack	Circumferential crack , Mixed crack, Irregular cracks, Rapid Crack Propagation (RCP)
11	Fracture	Circumferential fracture, Mixed fractures
12	Crazing	Through crazing, External crazing, Internal crazing
13	Deformed	Ring
14	Buckling	Axial buckling, Transverse/Ring buckling, Non-symmetric buckling, Longitudinal buckling, Other
15	Leakage/Burst	Snap through, Two-sided snap Through, perforation, Pinhole
16	Dimension Change	
17	Blisters/Voids	External blister, Internal blister, Blazes, Delamination
18	Abrasion/Erosion	
19	Chemical Attacks	
20	Collapse	
21	Other	Blank ends, Blown T, ball valve, End blown off
Operational and Maintenance Failure		
22	Obstacles/Obstructions	Object intruding through wall, Object wedged in the joint, Object through connection, External pipe or cable, Other obstacles
		B. Joint
Typical Failure Mode		
1	Push-fit and Slip-on Collar (Gasket)	Displaced gasket (into socket), Cracking of socket
Rare Failure Modes		
2	Push-fit and Slip-on Collar (Gasket)	Spigot deflection, Gasket degradation, Bell splitting, PVC material degradation
3	Solvent Weld	Poor adhesion degradation, embrittlement, Bell splitting, Angular deflection, Misalignment
4	Other	Joint leak, Mechanical joint leak, Blown rubber on flange, Blown joint, Leak on collar, Leaking rubber ring, Gasket/Seal failure, Structural bell failure, Butt fusion joint, Electro-fusion joint, Flange joint failure

Table 3-20. Failure Modes of Steel Wastewater Pipe (Force Main).

Steel Wastewater Pipe Failure Modes (Force Main)	
	A. Main Line
Typical Failure Mode	
Structural Failure	
1. Hole	Blowout holes
2. Corrosion	External corrosion, Internal corrosion
3. Deformed	Vertical deformed, Horizontal deformed
4. Lining Failure	Detached, Defective end, Blistered, Service cut shifted, Abandoned connection, Buckled, Wrinkled, Spall, Crack
5. Burst Failure	Burst failure (int.), Burst failure (ext.), Burst failure (3rd party)
6. Weld Failure	Longitudinal, Circumferential, Multiple, Spiral
Operational and Maintenance Failure	
7. Exfiltration	
Rare Failure Mode	
Structural Failure	
8. Crack	Circumferential crack, Longitudinal crack, Spiral crack, Mixed crack
9. Fracture	Circumferential fracture, Longitudinal fracture, Spiral fracture, Mixed fracture, Shearing, Bell shearing
10. Collapsed	
Operational and Maintenance Failure	
11. Obstacles/Obstructions	Object intruding through wall, Object wedged in the Joint, Object through connection, External pipe or cable, Other obstacles
	B. Joint
Failure Mode	
Typical Failure Modes	
1. Push-On	Gasket/seal failure, Angular deflection, Axial pull out, Thermal expansion/cracked spigot, Lateral pipe movement, Internal corrosion, External corrosion
2. Bolted Flange	Bolt failure, Flange failure, Gasket failure, Stress at flange, Bolt corrosion
Rare Failure Modes	
3.Mechanical joint	Bolt failure, Bolt corrosion, Backing ring corrosion, Structural spigot failure
4.Slip-On Coupler	Internal corrosion, External corrosion, Bolt corrosion, Spigot pull-out

Wastewater pipes failure modes and mechanisms of each material type including cast iron, ductile iron, PVC, polyethylene, pre-stressed concrete cylinder, and clay are explained subsequently.

3.2.1 Concrete Wastewater Pipe Failure Modes and Mechanisms

Most of the time, failure of concrete pipe occurs as a result of deflection and cracking related to excessive external loading and corrosion. Initial failure is possible and usually caused by design error or impact during construction. Corrosion leads to structural failure. Sulfuric acid that corrodes the pipe near the crown is formed by hydrogen sulfide gas (H_2S) in sewage with condensing fluid [52]. Pre-stressed concrete cylinder pipe is constructed by circumferentially winding a high strength steel wire, under tension, around a concrete core containing a sheet steel cylinder, which serves as a watertight membrane. Typically, ability of the steel pre-stressing wire determines the service life of such pipe [53]. Concrete is a mineral-based material. Softening of the pipes can occur through lime leaching when certain soil and water types are contacted. The process also involves corrosion in reinforcing wire with a corresponding loss of the structural integrity as the physical strength of the wires degrades. The reinforcing/pre-stressing wire in reinforced concrete pipe/pre-stressed concrete cylinder pipe is susceptible to stray current corrosion if the cathodic protection or other corrosion resisting components are not present. If the pre-stressing wire corrodes and swells, the mortar may crack. This can be responsible for catastrophic failure of pipe wall structure [54].

3.2.2 Ductile Iron Wastewater Pipe Failure Modes and Mechanisms

In most cases, corrosion leads to failure in ductile iron wastewater pipes. Corrosion is either internal or external. Internal corrosion is caused by exposure to highly corrosive substances in wastewater from industrial sources. Rarely, domestic wastewater by itself is corrosive enough to cause pipe failure.

In gravity pipelines, slime is built up and consequently results in corrosion if the wastewater flow is very slow (velocity less that 2 ft/s). In comparison to water pipes, wastewater force main pipelines are unlikely to have induced strain from operation pressure because of the lower pressure used in the system.

External corrosion occurs when the pipeline is subjected to a corrosive environment. The corrosion rate can be determined by the characteristics of the surrounding soil. Resistivity, redox potential and pH level are believed to be the predominant factors. Corrosion usually results in localized attack and formation of corrosion pits [54]. These pitting holes are unlikely to cause pipe structural failure however; they lead to leakage, infiltration, and exfiltration.

Designs for corrosion protection are very important in ductile iron pipe. Internally, the pipe is typically protected from corrosion by cement mortar. Externally, the pipe is commonly shielded by factory-applied zinc coating. One of the other issues in corrosion deterioration is stray current. Stray currents affecting underground pipelines are direct currents flowing through the earth.

3.2.3 Cast Iron Wastewater Pipe Failure Modes and Mechanisms

Cast iron is made by adding larger amounts of carbon to molten iron than would be used to make steel. Cast iron typically has 2.5-4% carbon by weight whereas steel has less than 1.2% carbon by weight. Adding more carbon makes it more fluid and easier to cast; however, when the metal solidifies, some or most of the carbon may form graphite flakes [49, 55].

Cast iron was manufactured using two methods. The older pipes were made of pit casting, a static process. And the newer pipes were made of centrifugally spun casting. Fundamentally, cast iron is brittle material that has less yield strength than ductile iron.

Main categories of failure that occur in cast iron pipes are cracks, which lead to fractures and collapses, caused by internal or external loading. The predominant pipe stresses in smaller pipe (less than 12 inches diameter) are caused by externally induced bending loads and can cause failure in small diameter pipe. In larger pipe, crushing failures predominate leading to longitudinal factures [54]. Two main forms of corrosion are general corrosion and localized pitting corrosion. Losses in pipe wall thickness resulting from corrosion combined with the external loads can initiate failure. Also, presence of the pitting corrosion can form holes.

When stray electrical currents are attracted by a metallic pipeline, they can induce electrolytic corrosion of the metal or alloy [56]. Stray current protection is recommended if the pipeline is located close to the sources of stray current, which include cathodic protection systems, direct current power trains or street cars, arc-welding equipment, direct current transmission systems, and electrical grounding systems.

In small pipe (less than 12 inches in diameter), the pipe stresses may be externally induced bending loads and leading to failure even where there is negligible corrosion. However in the larger diameter pipes, crushing failure is dominate resulting in longitudinal fractures [54]. These type failures once predominantly in cast iron pipes.

3.2.4 Polyvinyl Chloride (PVC) Wastewater Pipe Failure Modes and Mechanisms

PVC pipes are made of viscous-elastic material. Creep and stress relaxation are considered in this material type. Generally, PVC pipes are stronger in circumferential direction than longitudinal direction because of the oriented structure of the pipe material. Conversely, longitudinal-oriented PVC pipe exists but is rare. Axial over deflection induces bending stress along the pipe and eventually causes cracking. PVC pipe failure also can be caused by cyclic fatigues. Cyclic loading from turning pumps on and off can lead to premature failure in PVC pipe.

During construction or transit, accidents may cut or damage the pipe surfaces both internally and externally. Poor installation and workmanship may result in damage to the pipe. For example, over insertion may crack the end of the pipe.

3.2.5 Polyethylene and High-Density Polyethylene Pipe Failure Modes and Mechanisms

Polyethylene and high-density polyethylene are viscous-elastic material. Creep and stress relaxation should be considered in these material type. Various factors affecting polyethylene pipe include but are not limited to chemical attack, loading, temperature, and installation practices.

Three basic modes of polyethylene pipe failure are recognized as ductile failure at relatively high stresses, brittle fracture at intermediate stress levels, and environmental stress cracking or stress corrosion cracking at low stress levels [57].

3.2.6 Clay Wastewater Pipe Failure Modes and Mechanisms

Clay or vitrified clay pipes are strong and chemical resistant because of their ceramic material properties. Crushing failure rarely occurs in new clay pipe however, clay pipes are brittle. The properties of the clay pipe do not change over time so age of the pipe does not affect the durability of the pipe. The typical structural failure mode of the clay pipe is cracks. Failure in clay pipes is caused by three factors, loading, bedding system, and foundation.

Most of the failures in clay pipe resulting from construction, bedding, or foundation become evident in the first two years as soil consolidates completely. During construction, it is very important that the bell holes are dug properly or else the pipe bell may crack. Localized failure is possible as a result of inadequate longitudinal support, poor bedding, or overcompacting. Excessive point loading can create a star break in clay pipe. Impact failure resulting from third-party damage is feasible as well. Poor construction may lead to lateral shear, and differential settlement may cause shear between the manhole and the pipe.

Joints are the key weakness in clay pipes. Parallel offsets, pulled joints, excessive angular deflection, and leakage are the main joint failure modes of clay pipes. These joint failures may lead to root intrusions and infiltration, and inflow (I&I), which eventually affect the hydraulic capacity and functionality of the pipe.

3.2.7 Brick Sewer Failure Modes and Mechanisms

Failure modes of brick sewer consist of mortar loss caused by abrasion, poor cleaning practice and chemical attack, and cracking due to excessive external loading, and differential settlement. The failure of mortar will eventually lead to infiltration and exfiltration. This infiltration will carry soil particles into the sewer leading to loss of bedding support [52]. Other failures include root intrusion, sediment blocking the flow and object intrusion.

All failure modes and mechanisms of wastewater pipes by material are summarized in Appendix B. Table 3-21 is a list of the attached Microsoft Excel worksheets.

Table 3-21. List of Attachment for Wastewater Pipe Failure Modes and Mechanisms Based on Material Type [8, 13, 15, 36, 40–57].

Attachment	Appendix Material	System Type
B-1	Concrete pipe	Gravity
B-2	Reinforced concrete pipe	Gravity
B-3	Clay pipe	Gravity
B-4	Ductile iron pipe	Gravity
B-5	Cast iron pipe	Gravity
B-6	Polyvinyl chloride (PVC) pipe	Gravity
B-7	Polyethylene and high-density polyethylene pipes	Gravity
B-8	Brick	Gravity
B-9	Pre-stressed concrete cylinder pipe	Force Main
B-10	Ductile iron pipe	Force Main
B-11	Cast iron pipe	Force Main
B-12	Polyvinyl chloride pipe	Force Main
B-13	Steel pipe	Force Main

CHAPTER 4.0

DATA STRUCTURE FOR LIKELIHOOD OF PIPE FAILURE

Chapter Highlights

This chapter presents the development of standard data structure for pipe performance evaluation and description of data collection protocols and methods:

- **Objective:** Standard data structure can be used in development of pipe condition index, prediction model, and prioritizing repair and rehabilitation.
- **Data Structures:** Wood, Bronze, Silver, and Gold standard data structures are proposed for different size of the cities and work force in data collection.
- **Data Classes:** Parameters were divided into five classes based on their characteristics: Physical/Structural, Operational/Functional, Environmental, Financial, and Others.
- **Data Confidence:** All collected data will be classified as Direct Data, Derived Data, and/or Educated Guess Data.
- **Data Storage:** Wastewater pipe infrastructure information is frequently recorded in several cities and utilities' documents such as maps, maintenance records, daily field logs, recorded flow data, as-built (or record) drawing, specification, and survey information.
- **Data Compilation:** Often, the compilation of wastewater pipe infrastructure system data/information fails due to the lack of proper methodologies to easily retrieve the required data/information for analysis, and also the lack of integration between sections or departments within the utilities.
- **Data Collection Methods and Protocols:** The research team summarized data collection methods and protocols for every parameter in the data structure.

4.1 Development of Data Structure

The research team identified close to 100 possible parameters that may affect wastewater pipe infrastructure. The list of parameters was sent to participating utilities in the United States and abroad for comments and suggestions for improving the data structures. From the responses, changes and updates have been made. The goal is to eventually create a national standard data structure for wastewater pipe infrastructure. Along with an inclusive list of all parameters that ultimately can have an effect on the deterioration of the pipe infrastructure, a supplemental goal is to generate a report detailing methods and protocols (See Section 4.2).

This standard data structure was developed to aid the decision making process for condition assessment in an asset management program. In addition, the data structure can be used for developing a condition index, a prediction model, prioritizing repair and rehabilitation, prioritizing inspection, planning operation and maintenance, developing a capital improvement program, and other decisions.

The parameters were divided into five classes based on their characteristics: physical/structural, operational/functional, environmental, financial, and others. These classes are presented in Figure 4-1.

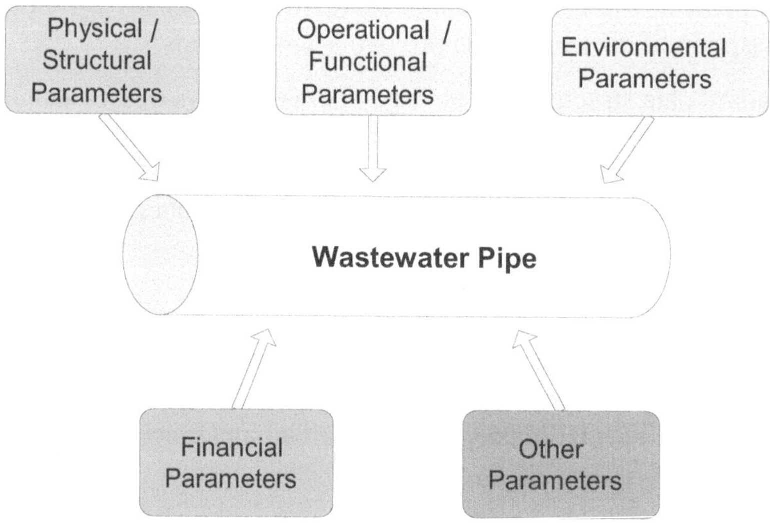

Figure 4-1. Classification of Pipe Parameters.

In this study, the research team considered data from major cities as well as smaller towns within the United States. The data structure was broken into four separate standards: Wood, Bronze, Silver, and Gold. Also, because the utilities lack readily available parameters, the wastewater pipe data structure was broken down into essential and preferable data.

Wood standard is for utilities that are very small and do not have sufficient resources to collect a large amount of data, or utilities that are establishing a wastewater pipe database. This standard represents the most important six essential and five preferable parameters for a wastewater pipe based on the research teams' knowledge of pipe infrastructure.

Bronze standard is for utilities that are small and do not have a lot of man-hours because of a small workforce. This standard represents 11 essential and 12 preferable parameters for a wastewater pipe.

Silver standard is for utilities that are larger than Bronze utilities; however, still may not have a special team designed to devote all their time to developing wastewater pipe data parameters within a system. This standard represents 23 essential and 28 preferable parameters for a wastewater pipe.

Gold standard is for the utilities that represent some of the largest cities within the United States and have a special team devoted to the continually update the wastewater database. This standard has 51 essential and 47 preferable parameters that affect wastewater pipes.

Ultimately, for the purpose of this study, most of the parameters are needed to develop a reliable prediction model and it is critical that the utilities provide the essential data. Because of time constraints of the utilities, the preferable parameters may be acquired by other means such as the research team organizing and researching past documents within the city data, or deriving from other sources. For example, utilities may not have soil parameters for a wastewater data structure; therefore, the research team may evaluate and search for from other sources such as Soil Survey Geographic Database, for that particular area.

Table 4-1 shows standard wastewater pipe data structure including Wood, Bronze, Silver, and Gold standards.

Table 4-1. Standard Wastewater Pipe Data Structure.

No.	Parameter	Unit	Gold	Silver	Bronze	Wood
Physical/Structural						
1	Node Identification Number (MH)	Node	E	E	E	E
2	Pipe Material	Type	E	E	E	E
3	Pipe Diameter	Inch	E	E	E	E
4	Pipe Age	Year	E	E	E	E
5	Pipe Depth	Feet	E	E	E	P
6	Node Length	Feet	E	E	P	
7	Pipe Wall Thickness	Inch	E	P		
8	Pipe Location	Area	E	E	P	
9	Pipe Shape	Type	E	E	P	
10	Pipe Joint Type	Type	E	E	E	P
11	Function of pipe	Type	E	E	E	E
12	Pipe Bedding	Yes/No-Type	E	E	P	
13	Trench Backfill	Type	E	E	P	
14	Construction Specification	Spec.	E	P		
15	Pipe Slope	Gradient	E	E	P	
16	Pipe Design Life	Year	E	P		
17	Pipe Design Strength	psi	E	P		
18	Pipe Lining	Yes/No-Type	E	E	P	
19	Manhole Condition	Record	E	P		
20	Pipe Quality	Level	P			
21	Pipe Section Length	Feet	P			
22	Pipe Vintage	Year	P			
23	Pipe Lateral	Type	P			
24	Dissimilar Materials	Yes/No	P			
25	Pipe Installation	Rating	P			
26	Pipe Manufacture	Record	P			
27	Pipe Trench Width	Feet	P			

No.	Parameter	Unit	Gold	Silver	Bronze	Wood
28	Pipe external Coating	Yes/No - Type	P			
29	Pipe Cathodic Protection	Yes/No - Type	P			
30	Pipe Thrust Restraint	Yes/No - Type	P			
31	Lateral Connections	Record	P			
32	Pumping Station and WWTP	Record	P			
Operational/Functional						
33	Wastewater Quality	Record	E	P		
34	Wastewater Pressure (Force Main)	psi	E	P		
35	Pipe Hydraulics	Gallon/Min	E	P		
36	Pipe Surcharging	Yes/No - Ft.	E	P		
37	Operational & Maintenance Practices	Type- Level	E	E	P	
38	Pipe Renewal Record	Record	E	E	E	P
39	Pipe Failure Record	Record	E	E	E	P
40	Infiltration/Inflow	Level-Gal/Min	E	E	P	
41	Exfiltration	Level	E	P		
42	Blockage/Stoppage	Yes/No-Type	E	P		
43	Sediments	Ton/Feet	E	P		
44	Inspection Record	Yes/No-Record	E	P		
45	Flow Velocity	Feet/Second	E	P		
46	Sewer Odors	Yes/No	P			
47	Sewer Flooding	Yes/No	P			
48	Sewer Overflow (SSO/CSO)	Yes/No	P			
49	Backup Flooding	Number	P			
50	Leakage Allowance	%	P			
51	Interruptions	Record	P			
Environmental						
52	Soil Type	Type	E	E	E	E
53	Soil Corrosivity	Level	E	P		
54	Soil Resistivity	Level	E	P		
55	Redox Potential	Level, mV	E	P		
56	Soil Moisture Content	Percent	E	P		
57	Stray Currents	Yes/No	E	P		
58	Groundwater Table	Feet	E	P		
59	Ground Cover	Type	E	P		
60	Loading Condition (Dead Load)	Lbs/sq.ft.	E	E	P	
61	Loading Condition (Live Load)	Avg. daily traffic-level	E	E	P	
62	Rainfall/Precipitation	Inch/year	E	P		
63	Climate - Temperature	^0F	E	E	P	
64	Topography	Map	E	P		
65	Extreme Events	Yes/No - Type	E	P		
66	Soil Disturbance	Yes/No	P			
67	Runoff Rate	Cu. Ft/Sec.	P			

No.	Parameter	Unit	Gold	Silver	Bronze	Wood
68	Non-Uniform Soil	Yes/No	P			
69	Frost Penetration	Yes/No-depth	P			
70	Non-Uniform Slope	Yes/No	P			
71	Unstable Slope	Yes/No	P			
72	Seismic Activity	Yes/No	P			
73	Catchment Area (Sewershed)	Sq.Ft	P			
74	Average Closeness to Trees	Feet	P			
75	Tidal Influences	Yes/No	P			
76	Soil pH	pH	P			
77	Soil Chloride	%	P			
78	Soil Sulfate	%	P			
79	Soil Sulfide	%	P			
80	Pipe Connections	Type	P			
Financial						
81	Annual Capital Cost	$/Year	E	E	E	P
82	Annual Maintenance Cost	$/Year	P			
83	Annual Repair/Rehabilitation Cost	$/Year	P			
84	Installed and Replacement Cost	$	P			
85	Annual Operational Cost	$/Year	P			
86	Annual Energy Cost	$/Year	P			
87	Depreciated Value	%	P			
88	Benefit/Cost	Record	P			
Others						
89	Customer Complaint	Type	E	P		
90	Chemical	Event	E	P		
91	FOG	Yes/No	E	P		
92	Overall Pipe Condition	Rating	E	P		
93	Density of Connections	Number/ Mile	P			
94	Resident Population Served	Number	P			
95	Failing Utilities	Yes/No	P			
96	Consequence/Risk	Level	P			
97	Third-party Damage	Yes/No	P			
98	Other Information	-	P			

*** Note: E: Essential Parameters, P: Preferable Parameters**

Both essential and preferable parameters were developed for the wood, bronze, silver, and gold wastewater infrastructure system. The parameters are set up as listing each parameter into one of the five classes as shown in Figure 4-1 and also provide the parameter units. A brief explanation is provided for each parameter to supply an additional description. The confidence of data is also listed within the spreadsheet for the utilities to provide to the research team. This space is provided for the research team to know the confidence of the data or parameters supplied by the utilities. The confidence level is separated into three different classes: direct record, derived indirectly, and educated guess.

The wood, bronze, silver, and gold standard wastewater data structures are presented in detail in Appendix D. Each of the parameters included within the gold standard data structure are explained in detail in Appendix E. There are two separate documents listing the detailed parameters: Gold essential parameters and gold preferable parameters. Table 4-2 lists the attached documents.

Table 4-2. List of Attachments for Wood, Bronze, Silver, and Gold Wastewater Data Structures and Gold Essential and Preferable Wastewater Data Structure Parameters. [7, 8, 35-38, 43, 45, 48, 49, 58–72]

Attachment	Appendix Material
C-1	Wood Data Structure (Essential & Preferable)
C-2	Bronze Data Structure (Essential & Preferable)
C-3	Silver Data Structure (Essential & Preferable)
C-4	Gold Data Structure (Essential & Preferable)
C-5	Gold Essential and Preferable Parameters Detail Description

4.2 Description of Data Collection Methods and Protocols

After developing the standard data structure for the wastewater pipe infrastructure presented in Table 4-1, the research team identified data collection methods and protocols for quantifying identified parameters. The wastewater pipeline data can be collected by direct or indirect methods, for example, direct methods such as various inspection methods, and indirect methods such as hydraulic modeling.

Wastewater pipe infrastructure information frequently has been recorded in several cities and utilities' documents such as maps, maintenance records, daily field logs, recorded flow data, as-built (or record) drawing, specification, and survey information. These documents may have been maintained by different city departments. Often, the compilation of all wastewater pipe system information fails because of the lack of proper means to store the data in an easily retrieved format and the lack of integration among sections or departments. The most effective format would be an electronic overall base map such as aGIS for data storage. In addition to cities and utilities' documents mentioned above, staff interviews, survey questionnaires, and informal meetings are always good sources for information.

However, if the information needed for processing is not available, the utilities might need to collect missing data. The data collection is expensive and requires extra workforce and time, so selecting an appropriate inspection technique for collecting each type of data is necessary. For example, one of the most widely used inspection techniques for collecting internal wastewater pipeline information is called closed-circuit television (CCTV). From CCTV images the defects in the pipe such as cracks, fractures and holes are captured. Root intrusions through the cracked or broken pipe, I&I and exfiltration can be detected via this technique as well. Other effective inspection techniques include smoke test, dye test, and manhole inspection.

The research team developed pipe data collection methods and protocols for every parameter in the data structure. Extensive study was conducted to provide a dependable list of pipe data collection methods and protocols. The data collection methods and protocols are defined into two categories, Sources and Inspection/Testing. Sources show the locations where the data usually stored. The examples of data collection methods and protocols are shown in the Figure 4-2. This example shows that the pipe location parameter can be found in the manhole survey, GIS, as-build drawing, specifications, and inspection records. In addition, it lists different inspection techniques and references that can be employed to acquire the needed data.

8. Pipe Location

Sources: Manhole survey, GIS, as-build drawing, specifications, inspection records

Inspection/Testing:

- Radio Frequency Transponder
- Magnetics
- Electromagnetic
- Acoustic reflectometry and ground probing radar
- Test pit

Ref:

- Manual Manhole Inspection and Rehabilitation, American Society of Civil Engineers (ASCE), Engineering Practice No. 92
- Inspection Guidelines for Ferrous and Non-Ferrous Force Mains, WERF, Jason Consultants LLC, 2007

Figure 4-2. Example of Data Collection Methods and Protocols – Pipe Location.

Other sources or databases are also useful, for instance average daily traffic from the department of transportation can be used to calculate live load pressing on pipes. Additional resources such as suggested websites are listed as well, if available. Figure 4-3 illustrates websites that can be accessed to download topography data.

45. Topography

Sources:

- NM Bureau of Geology and Mineral Resources- http://geoinfo.nmt.edu/publications/maps/topographic/home.html
- TerraServer.com http://www.terraserver.com/
- TopoZone.com http://www.topozone.com
- USGS Earth Resources Observation and Science (EROS) http://edc.usgs.gov

Figure 4-3. Example of Data Collection Methods and Protocols – Topography.

Inspection/Testing provides the literature sources related to specific testing method, if the utility decides to collect the data. Lab testing and experimental procedures from standard organizations such as American Society for Testing and Materials (ASTM), American Association of State Highway and Transportation Officials (AASHTO) are also recommended in this section, as shown in Figure 4-4.

33. Soil Type

Sources: GIS, other databases

Testing:

- American Society for Testing and Materials: ASTM D422 - 63(2007) Standard Test Method for Particle-Size Analysis of Soils
- American Society for Testing and Materials: ASTM D3282-93(2004)e1 Standard Practice for Classification of Soils and Soil-Aggregate Mixtures for Highway Construction Purposes
- American Association of State and Highway Transportation Officials : AASHTO M 145 Standard Specifications for Classification of Soils and Soil-Aggregate Mixtures for Highway Construction Purposes

Figure 4-4. Example of Data Collection Methods and Protocols – Soil Type.

The data collection methods and protocols are summarized in detail and presented in Appendix E. There are two separate documents listing the data collection methods and protocols; essential parameters data collection methods and protocols, and preferable parameters data collection methods and protocols. Table 4-3 lists the attached documents.

Table 4-3. List of Attachment for Data Collection Methods and Protocols. [7, 29, 28, 34, 36, 38, 56, 67, 74–88]

Attachment	Appendix Material
D-1	Gold Standard Essential Parameters Data Collection Methods and Protocols
D-2	Gold Standard Preferable Parameters Data Collection Methods and Protocols

CHAPTER 5.0

DATA STRUCTURE FOR CONSEQUENCE OF PIPE FAILURE

Chapter Highlights

This chapter presents a description of risk assessment and management and development of data structure for consequence of pipe failure:

- **Risk/Performance:** The risk/performance of the system is determined by two basic measures:
  1. Event: The probability of failures
  2. Consequence: The impact of failures

- **Risk Assessment:** Each risk event has its causes and identifying and understanding the causes will help in estimating and mitigating risk associated with breaches or failure of wastewater infrastructure systems.

- **Failure Consequences:** Seven impacts of failures for wastewater pipe have been identified: environmental, traffic flow, service disruption, time, financial, financial on private property, and other impacts.

- **Complexity/Classification:** Complexity of repair/rehabilitation/renewal, and pipe classifications are also discussed in this chapter.

5.1 Risk Assessment and Management

Risk is defined as: "The chance of something happening that will impact upon objectives, and is measured in terms of a combination of the likelihood and consequences of events" [83].

The risk/performance of the system is determined by two basic measures:

- Event: The probability of failures
- Consequence: The result of failures

The identification, assessment and control of risk are a key focus at all levels of asset management processes. Risk assessment is an essential element of risk management. Figure 5-1 shows a typical risk assessment diagram.

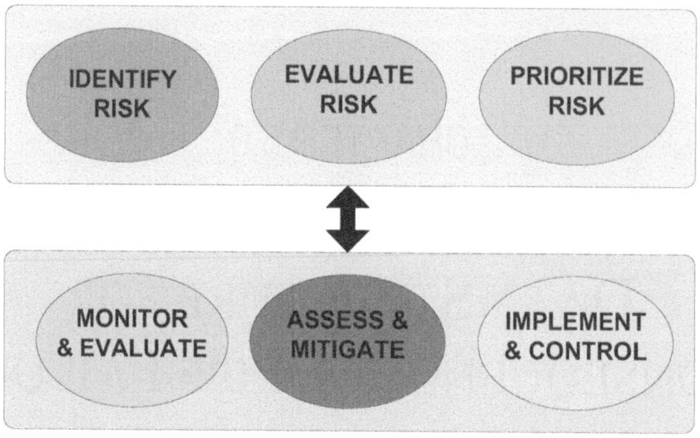

Figure 5-1. Typical Risk Assessment Diagram for Water Infrastructure.

5.1.1 Major Steps in Risk Assessment and Management

Risk is often specified in terms of events and consequences. An important step in the risk assessment process is to identify events and their possible causes. Each risk event has its causes and identifying and understanding the causes will help in estimating and mitigating risk associated with breaches or failure of wastewater infrastructure system.

5.1.1.1 Assess Likelihood of Events

There are many ways in which the likelihood of each event can be estimated. If reliable statistical records of system performance have been kept over a long period and conditions have either remained constant or followed steady trends, then predictions of future performance with associated probabilities can be made using the associated data. Standard probability theory can be used to manipulate the data. Alternatively, where the system is well defined but perhaps there is little data on its performance, the likelihood of a particular event can be calculated using Event Trees or Bayesian Belief Network, based on knowledge of the likelihood of failure of wastewater infrastructure system.

5.1.1.2 Assess Consequence of Events

Consequences of events can be evaluated from real life observations, tests, models or opinions. Whatever the means of evaluation, a good understanding of the system, its components and interactions is essential. Many models exist that can be used to calculate aspects of system performance, given various input configurations. For more complex system models, techniques such as fuzzy logic or conditional probability have been used.

5.1.1.3 Likelihood Rating Criteria

The analysis of risks requires an objective assessment of their frequency of occurring, based on historical events and assessment of what has changed and may occur into the future. An example of a typical scale used to rate the likelihood of different risks occurring is shown in Table 5-1.

Table 5-1. Example of Likelihood Rating Matrix.

Rating	Description	Probability	Frequency
1. Certain	Risk will certainly occur	XX	yy
2. Likely	Risk will likely occur	XX	yy
3. Possible	Risk may possibly occur	XX	yy
4. Unlikely	Risk will not likely to occur	XX	yy
5. Rare	Risk will rarely occur	XX	yy

5.1.1.4 Consequence Rating Criteria

In a qualitative risk assessment using the matrix method, scales are used to describe the magnitude of the consequence. An example of a possible qualitative scale to rate the consequence of different risks occurring is shown in Table 5-2.

Table 5-2. Example Consequence Rating Matrix.

Rating		Criteria				
		Health	Environment	Reputation	Regulation	Financial
A.	Extreme	A1	A2	A3	A4	A5
B.	Major	B1	B2	B3	B4	B5
C.	Moderate	C1	C2	C3	C4	C5
D.	Minor	D1	D2	D3	D4	D5
E.	Insignificant	E1	E2	E3	E4	E5

5.1.1.5 Risk Rating Criteria

Risk levels are then displayed in a matrix that applies rules to combine likelihood and consequence into a level of risk. In the example shown in Table 5-3, the outcomes of assessment are separated into five categories; that is, 5) extremely high risk, 4) high risk, 3) moderate risk, 2) low risk, and 1) extremely low risk.

Table 5-3. Example of Risk Rating Matrix.

	Likelihood of Failure				
Consequence of Failure	1	2	3	4	5
A	Medium	Medium	High	Extreme	Extreme
B	Medium	Medium	High	Extreme	Extreme
C	Medium	Medium	High	High	High
D	Low	Low	Medium	Medium	Medium
E	Low	Low	Medium	Medium	Medium

5.2 Development of Data Structure

A system is always considered to be in relation to its environment. The environment can be defined as a set of parameters that can affect the system or on the contrary, be affected by the system. The system in this case is the wastewater pipe, which can often fail and have environment and societal impacts.

The research team developed a pipe failure impact, pipe characteristic, and pipe renewal complexity data structure. In the proposed failure impact data structure, seven impacts of failures for wastewater pipe have been categorized; environmental, traffic flow, service disruption, time, financial, financial on private property, and other impacts. Any of these seven impacts may result in serious consequences. For example consider the impact on traffic flow. A failed wastewater pipe under a road can ultimately cause the road to collapse. Once collapsed, vehicles may be damaged from the disaster and an alternative route may be needed. Another example is the impact to the environment. If a wastewater pipe leaks because of pipe defect such as a crack or fracture, the wastewater will contaminate the environment. Leakage occurring in the vicinity of water sources may cause water contamination and outbreak of diseases to the people who are consuming or using this water for everyday life. Thus, the consequences of failure are used to capture the criticality of the effect that will occur if the pipe fails.

Pipe characteristics are also important to consider when investigating failure consequences. These characteristics include the function of pipe, depth of pipe size of pipe, location of pipe, age of pipe, material of pipe, and other factors. Pipe characteristics can help to explain why a pipe failure may occur. For example, an old pipe, defined as 51 years or older, may display many cracks and failures. These failures could lead one to conclude that older pipes have a greater rate of cracks and failures in comparison to newer pipes. These failure consequences are fundamental in prioritizing the renewal investment, in addition to the overall rating of condition in wastewater infrastructure systems. The pipe sections that have high failure consequences and a poor condition rating (assumed to correlate to a high likelihood of failure) will be selected first in the renewal process. The complexity of renewal should also be considered. The complexity involves at least nine issues: access to pipe, utility density, type of utility, utility pattern, traffic volume, quality of utility record, availability of repair materials, contaminated soils, and other issues. For example consider the utility density. Depending on the area, multiple utilities involving municipal, energy, and communication may exist in close proximity. When considering the repair/rehabilitation/replacement of wastewater pipe, construction specifications and local utilities should be contacted to determine if other utilities might be disrupted during construction. Breaking or disrupting another utility may cause a disastrous situation. Failure consequence data structures, including data structure for pipe failure impact, data structure for pipe characteristic, and data structure for pipe renewal complexity are presented in Table 5-4 through 5-6, consecutively.

Table 5-4. Data Structure for Pipe Failure Impact.

Failure Impact Data Structure
1. Environmental Impact
a. Not applicable
b. Impacts on conservation/Protection zones
c. Quality impacts on surface waters and/or surroundings
d. Legal issues (agency fines/ non compliance issues)
e. Others
2. Traffic Flow Impact
a. Alternative route available
b. Limited alternative route
c. No alternative available
3. Service Disruption Impact
a. Residential areas, etc.
b. Industrial/ commercial (general manufacturing and heavy industry)
c. Business, offices, etc. (light industry)
d. Water critical commercial businesses (car washes, hair dressers, Laundromats, water process, manufacturing)
e. Hospital, school, etc.
4. Time Impact – Length of time out of service
a. Short term
b. Medium length
c. Long term shut down
5. Financial Impact – Carrying out the repair of pipe section
a. Low ($)
b. Moderate ($$)
c. High/very high ($$$)
6. Financial Impact on Private Property
Property damages caused by broken pipe should be measured as an impact as broken sewers can cause surcharging and sewer backups of sewage into basements. These are usually part of the third-party damage costs that a utility may have to pay through insurances.
a. Low
b. Medium
c. High
7. Other Impact Factors

Table 5-5. Data Structure for Pipe Characteristic.

Pipe Characteristic Data Structure
1. Function of Pipe
a. Force main (sewage main)
b. Gravity Sewer
c. Water Main
d. Combined Sewer
e. Stormwater
2. Depth of Pipe
a. Less than 10 feet
b. Between 10 and 25 feet
c. Greater than 25 feet
3. Size of Pipe
a. Less than 18 inches
b. Between 18 and 36 inches
c. Greater than 36 inches
d. Trunk Sewer and Transmission Water Pipes
4. Location of Pipe
a. Urban
b. Suburban
c. Rural
5. Age of Pipe
a. New (0 – 25 years)
b. Average (26 – 50 years)
c. Old (51 – and above)
6. Material of Pipe
a. Rigid (Concrete, Cast Iron, Steel, etc.)
b. Flexible: (PVE, High-density Polyethylene, etc.)
c. Brittle: (Clay, etc.)
d. Others
7. Other Factors

Table 5-6. Data Structure for Pipe Renewal Complexity.

Renewal Complexity Data Structure (Constructability Issues)

1. Access to Pipe

Access to utilities refers to the difficulty or ease of access to buried utilities that may be encountered on the project.

a. Access available b. Limited access available c. Restricted access: under railway, bridge, building, river, etc.

2. Utility Density

Density of utilities refers to the number of buried utilities that can be expected to be encountered on the project. If there are many utilities expected to be buried at the project site, more reliable data/information will be required to successfully locate those utilities.

a. Low (1 to 2 pipes in close proximity) b. Medium (3 to 5 pipes in close proximity) c. High (more than 5 in close proximity)

3. Type of Utility

Type of utilities refers to the various service types of buried utilities that can be expected to be encountered on the project. Utilities can be broadly divided into three different categories: (1) municipal, (2) energy, and (3) communication.

a. Less-Critical: Water, sewer, stormwater

b. Sub-Critical: Telephone, electric, television cable, etc.

c. Critical: Fiber-optic cable, gas, oil, high-voltage line, etc.

4. Utility Pattern

Pattern of utilities refers to the configuration of buried utilities that can be expected to be encountered on the project. Some areas may have a simple pattern that consists of a few parallel or crossing utilities, while other areas may have a complex pattern that consists of numerous parallel and crossing utilities.

a. Simple: One parallel and/or one crossing utility

b. Average: Two parallel and/or two crossing utilities

c. Complex: More than two parallel and/or crossing utilities

5. Traffic Volume

Traffic volume is the average daily traffic (ADT) volume for the project per lane. Any delay in the project in areas with higher traffic volume will result in greater travel delays.

a. Low (≤ 1,500 ADT per lane) b. Moderate (> 1,500 and ≤ 6,000 ADT per lane) c. High (> 6,000 ADT per lane)

6. Quality of Utility Record

Quality of utility record indicates the reliability of existing records on buried utilities. The availability of accurate historical utility records for the project can significantly reduce the potential for encountering unexpected underground utilities.

a. Good b. Fair c. Poor

7. Availability of Repair Materials

a. High (lots of material available so low risk in getting job done etc) b. Medium c. Low

8. Contaminated Soils

a. Not applicable b. Low/Medium: special treatment not required c. High: special treatment required

9. Other Issues

CHAPTER 6.0

STANDARDIZATION OF FIELD DATA STORAGE AND RETRIEVAL FRAMEWORK

Chapter Highlights

This chapter presents a geo-spatial database to support data and models and five utilities cases studies, as well as 11 database development steps and database security.

- **Database Development:** The overall framework of geo-database development is described in 11 steps starting from meeting with participating utilities, acquisition of data, development of geo-database, to development of web-application.

- **Pipe Infrastructure Data Standardization:** By standardizing the way data and information is entered, we can increase the usefulness of the database

- **Data Source:** The research team collected and compiled data from various participating utilities. The research team also collected and compiled data from various sources (i.e. Soil data from U.S. Geological Survey).

- **Data Format:** The utilities pipe infrastructure data was received in many formats. For example, Tabular, Text, GIS Layers, etc.

6.1 Development of Geo-Database

The data analysis and development is a very extensive process considering many different sources, formats, modeling, and results. The overall framework for collecting and analyzing pipe infrastructure data for research is described in the following steps.

6.1.1 Step 1: Meetings with Participating Utilities

Memorandums of Understanding were signed with all participating utilities and meetings were conducted to facilitate the data transfer process, as well as to discuss the current state of infrastructure data management in each community. Most meetings have been conference calls between the researchers and the communities' utility and GIS managers. In some of these meetings, the use of computer screen broadcasting software (in this case, Go-To-Meeting), has improved communication by allowing the meeting participants to show specific data and software remotely. Figure 6-1 is a photo taken during the Seattle net-meeting.

Figure 6-1. Net-Meeting with Seattle Public Utility, WA.

In general, the meeting process follows a standard order as shown in Table 6-1. After an introductory meeting, a subsequent data transfer meeting is usually required to negotiate the technicalities of how the data will be transferred. Once the data is received, the researchers may have follow-up questions regarding certain aspects of the data.

Table 6-1. Net-Meeting and Site Visit with Participating Utilities.

Participating Utilities	Introductory Meeting	Data Transfer Meeting	Data Received	Follow-up Questions
Seattle, WA	YES	YES	YES	YES
Orange County, CA	YES	YES	YES	YES
Blacksburg, VA	YES	YES	YES	YES
Pittsburgh, PA	YES	YES	YES	YES
Atlanta, GA	YES	YES	YES	YES

In addition to meetings with the communities, meetings have taken place with the San Diego Supercomputer Center (SDSC); these meetings were necessary to set up the database link between Virginia Tech and the SDSC.

6.1.2 Step 2: Utility Data Transfer

The methods for data transfer from each community varied with the preferences of each community, and with technical limitations. Originally, an FTP (file transfer protocol) site was set up to allow communities to send data to the researchers. However, the FTP file transfer appeared to be efficient only for relatively small data transfers; transfers of large files were requiring hours to complete, and in some cases, the transfer would fail part-way through. Instead of using FTP, in many cases it proved more convenient to transfer data on physical discs (CDs or DVDs). The DVDs from Seattle, Pittsburgh, Louisville, Orange County, and Atlanta were received, and initial analysis was done. DVDs provided a more convenient way to transfer data; however Orange County Utility preferred to send an FTP link. The data is downloaded from the server and now being analyzed, as shown in Figure 6-2.

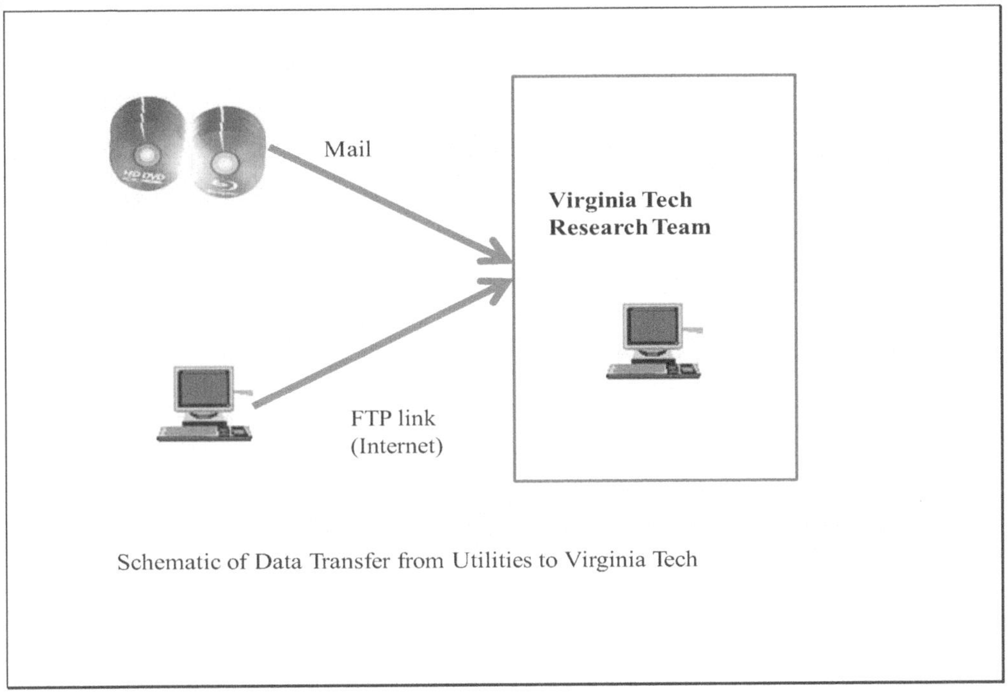

Figure 6-2. Utility Data Transfer Framework.

6.1.3 Step 3: ArcSDE Server for Geospatial and Tabular Data

After receiving the communities' GIS and tabular data, the researchers stored it at the San Diego Supercomputing facility, using an advanced spatial data server (ArcSDE) connection to Oracle table space at San Diego. The ArcSDE connection is being hosted by Virginia Tech's E-Corridors program.

Once loaded through ArcSDE, the geospatial and tabular data may be viewed directly in GIS software (with appropriate access permissions). The access permissions have been given to the research team only.

The data from Seattle was uploaded as a test trial to check for efficiency and speed of transfer. This trial proved to be a good test, with tabular data of around 30,000 pipes transferring in 2 hrs. Once the data is uploaded to the server, subsequent retrieval and viewing is very quick, almost as though it were hosted locally.

The proposed effective structure to store data is to create a feature dataset for each city by its name and then create individual feature classes for manholes, pipelines, etc. in each dataset.

The main advantage of this structure is that the data from each city is stored separately and can be retrieved also in a similar fashion.

The illustration in Figure 6-3 gives a better understanding of the structured storage. The table names (manholes, pipelines, laterals, hydrants, valves, etc.) may change according to the final proposed data model.

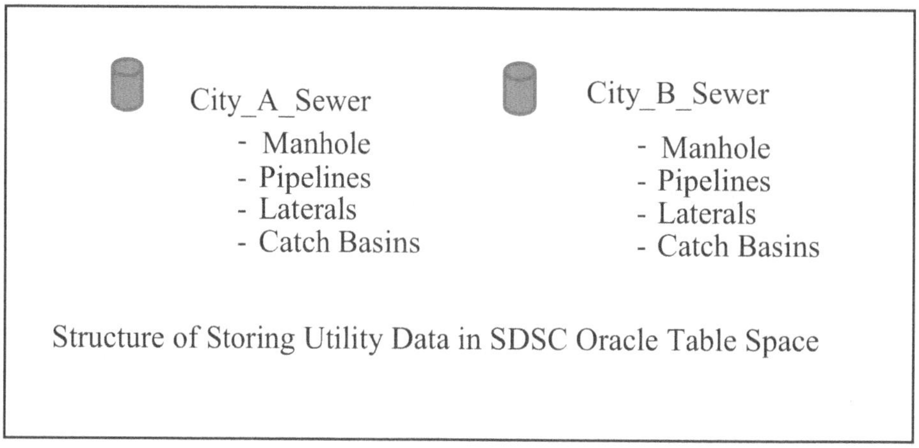

Figure 6-3. Storage of Utility Data in Standard Format.

6.1.4 Step 4: Uploading Utility Data for Analysis

The data received from utilities would be reviewed and initial analysis would be done to become familiar with their fields and attributes. This data will be then loaded to the standard data model template, this would further be transferred to the SDSC Oracle table space. The schematic in Figure 6-4 gives a better understanding of the data transfer to SDSC.

6.1.5 Step 5: Development of a Standard Geospatial Database

The research team will construct an example geospatial database of aggregated wastewater pipe infrastructure data. As illustrated in Figure 6-5, a common data framework to support condition assessment was defined and populated by collating disparate datasets from several partner utilities.

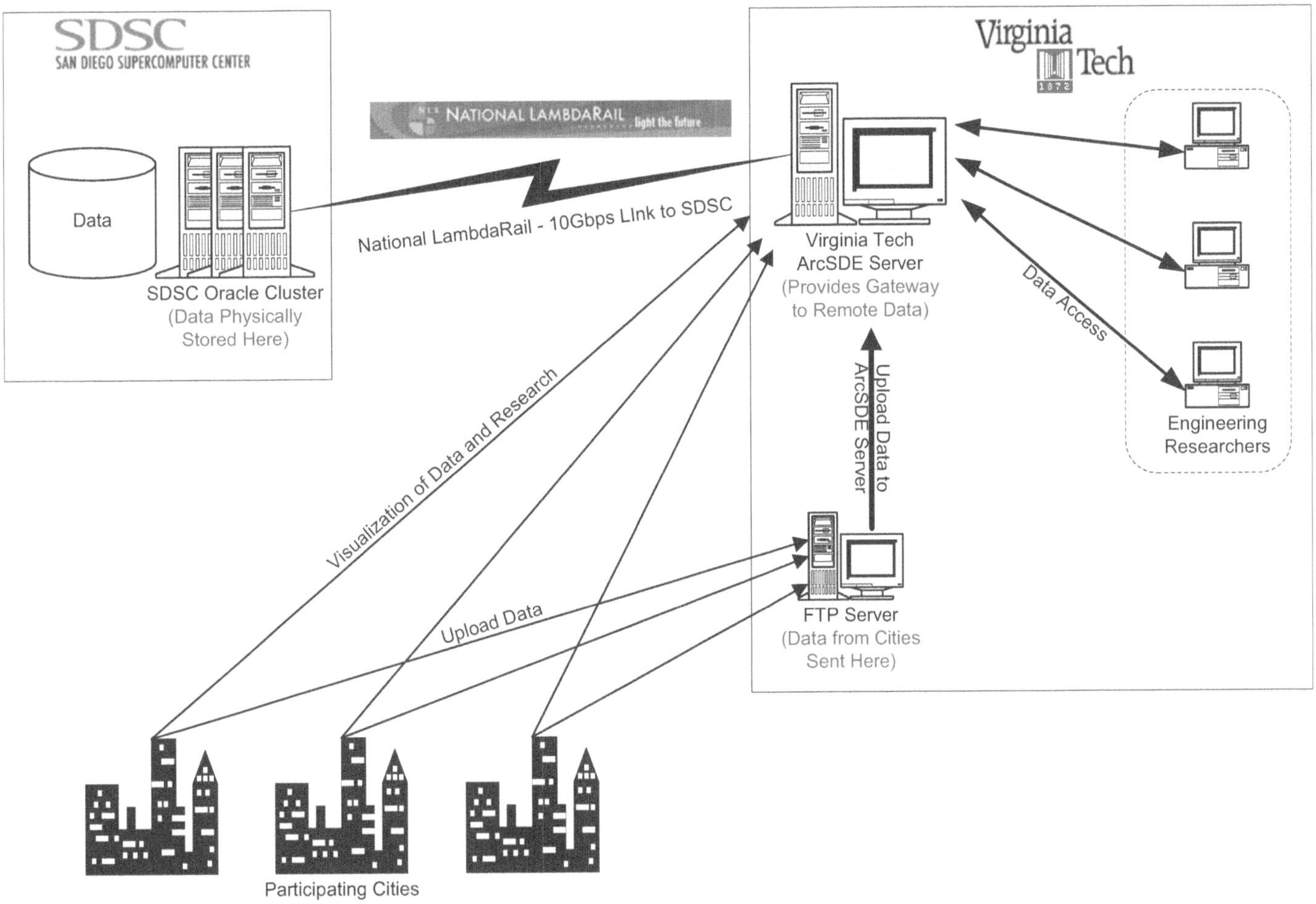

Figure 6-4. Data Transfer, Analysis, and Visualization Protocol.

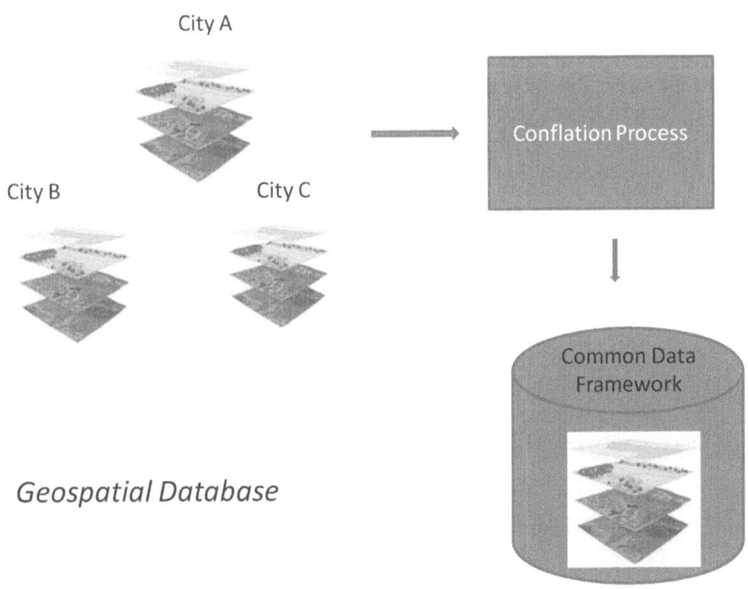

Figure 6-5. Conflation Process between Multiple Utilities.

6.1.6 Step 6: Development of Geo-Database Data Model

A geo-database data model is created from the standard data structure proposed by the research team. A screen shot of the data model is shown in Figure 6-6. This data model is used to develop mapping files. These files are used to transform the data from the utility files to the standard data model. The new geo-database is stored at the San Diego Supercomputing Center.

Feature Class: Nodes

Contains node (point) features in the sewer network, primarily manholes.

No.	Field Name	Alias	Description	Data Type	Allow Null?
1	ID		unique ID for this structure	String	No
2	Type		type of structure (e.g., manhole, junction, etc.)	String	yes
3	Material		structure material	String	yes
4	TopElev	Top Elevation	Rim elevation (or overflow elevation)	Double	yes
5	InvElev	Invert Elevation	Invert elevation of this node	Double	yes
6	Depth		Depth of structure (Rim - Invert elevation)	Double	yes
7	Shape		structure internal cross-sectional shape	String	yes
8	Dia	Diameter	diameter (if circular)	Double	yes
9	Dim	Dimensions	rectangular dimensions	String	yes
10	DateInst	Date installed	date of installation of the node	Date	yes
11	Status		operational status (active, abandoned, etc.)	String	yes
12	Operator		utility name or operator	String	yes
13	Source		link to record in source citation table	Long	yes

Figure 6-6. Geo-Database Data Model.

6.1.7 Step 7: Existing Geo-database – *a Priori* Data

There are several existing geo-databases relevant to pipe infrastructure that are available online or published. Some of these features included soil, geology, land cover, river, forcing, DEM, and stream flow. These geo-databases will help the research team in comparing the data provided from the utilities as well as filling in missing data parameters of the utility databases. For example, if the utility does not know the soil type within the area, the utility or the research team can visit online to the Conterminous United States, Soil Survey Geographic Database, or State Soil Geographic Database websites to determine the type of soil. A sample of existing geo-databases relevant to pipe infrastructure can be seen in Figure 6-7.

GeoDataBase - A Priori Data

Feature	Property	Source
Soil	Porosity, Sand, Silt, Clay Fractions, Bulk Density	CONUS, SSURGO and STATSGO http://www.soilinfo.psu.edu/index.cgi?soil_data&conus http://datagateway.nrcs.usda.gov/NextPage.asp http://www.ncgc.nrcs.usda.gov/products/datasets/statsgo
Geology	Bed Rock Depth, Kx/Kz	http://www.dcnr.state.pa.us/topogeo/, http://www.lias.psu.edu/soil/guides/X.html
Land Cover	LAI	http://glcf.umiacs.umd.edu/data/landcover/data.shtml, http://ldas.gsfc.nasa.gov/LDAS8th/MAPPED.VEG/LDASmapveg.shtml
	Manning's Roughness	Hernandez et. al., 2000
River	Topology	Derived using pihmGIS (Kumar et. al., 2007a)
	Manning's Roughness	Dingman (2002)
	Coefficient of Discharge	ModHms Manual
	Shape and Dimensions	Derived from regression using depth, width and discharge data from http://nwis.waterdata.usgs.gov/usa/nwis/measurements
Forcing	Precip Temp	Gauge data from MARFC. 6 hourly point data was spatially gridded conforming to the monthly precipitation distribution obtained from parameter-elevation regressions on independent slopes model (PRISM) (Daly et. al., 1994, 1997)
DEM		http://seamless.usgs.gov/
Streamflow		http://nwis.waterdata.usgs.gov/nwis/rt

Figure 6-7. Existing Geo-Database.

6.1.8 Step 8: Example of Geo-Database Visualization System

An example of geo-database layers and visualization system is shown in Figure 6-8.

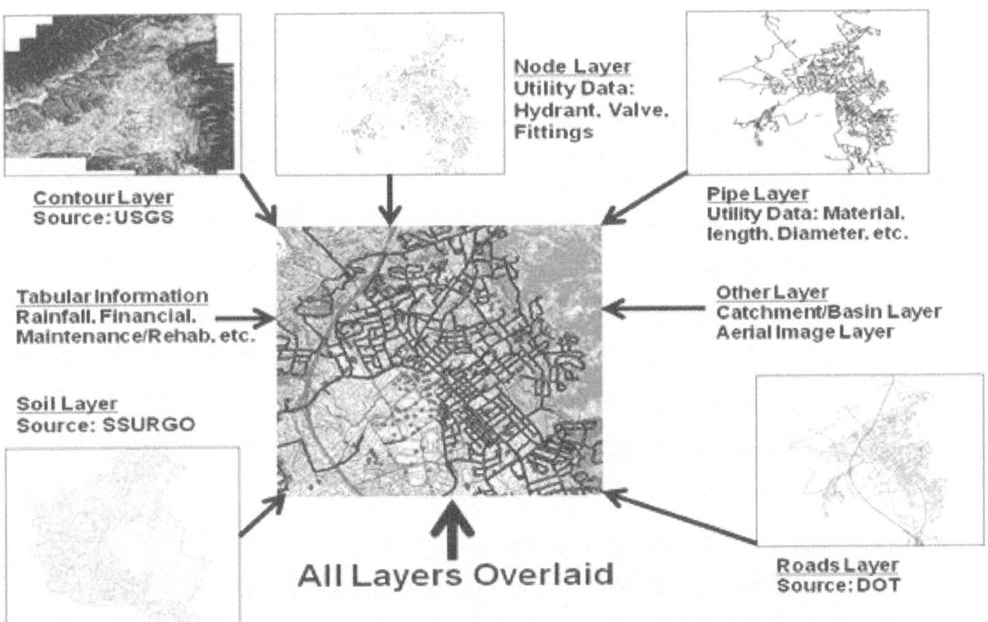

Figure 6-8. Example of Geo-Database Visualization System.

6.1.9 Step 9: Development of Web Application

A prototype web application was developed to query data with different privileges depending on the users, as shown in Figure 6-9. This application will help utilities and researchers to retrieve data with restricted access. Also, this might be used as an interface for the public to know more about utility programs and reports. The research team at Virginia Tech will have full access to query any kind of data.

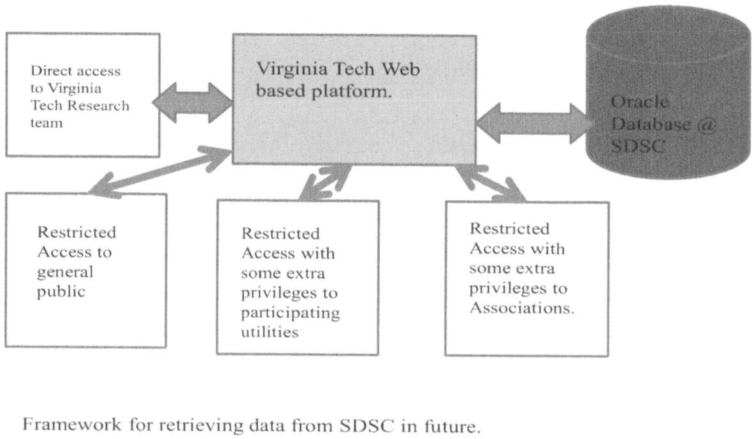

Figure 6-9. A Prototype Web Application Framework.

A prototype web application was developed to demonstrate how the aggregated data in the example geospatial database can be published, queried, and visualized in a map-based browser application (Figure 6-10). The application will deliver rich, interactive maps, data exploration tools, and reports through the use of Adobe Flash, a browser-independent vector-animation technology, and Geo-Server, an open-source web mapping server.

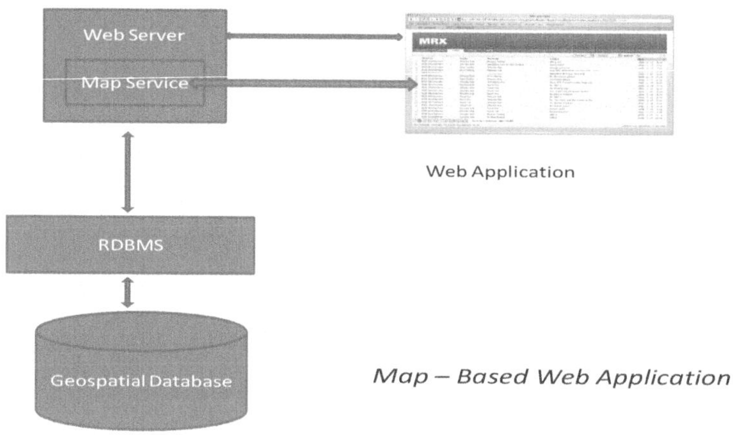

Figure 6-10. Map-Based Web Application and Supporting Components.

6.1.10 Step 10: Pipe Infrastructure Data Standardization

A standard is something set up and established by authority, custom, or general consent as a model or example. In the data world, a standard is a model to which all objects of the same class must conform and be represented by a standard definition. By standardizing the way data and information is entered, we can increase the usefulness of the database by having:

- More accurate data query results
- Better data mining and visualization
- Help in model development for decision making

Key issues related to standardization of the pipe data model are listed below:

- **Standard:** A data standard is a format representation for data values that can be described using a series of rules. In addition, in order for a format representation to be a standard, there must be an agreement among all participating parties that their data will conform to that standard.
- **Benefits:** Standardization is a process by which all elements in a data field (or set of related data fields) are forced to conform to an industry standard. There are many benefits to this process. The clear benefit is conformity for comparison and benchmarking.
- **Defining a Standard Model:** The important step in defining a standard is defining a data model, because a standard data model is not a standard until it is recognized as such through the concurrence of the users.
- **Testing the Standard Model:** If a data value is dictated to conform to a standard, there must be a way to test to see if a data value conforms to the standard data model.
- **Transforming into Standard Model:** It is important to create the ability to transform data into standard model from a non-standard source.
- **Updating the Database:** It is important to create the mechanism and procedure to update the standard database at the regular interval of time based on need and cost involved in collecting the data.
- **Maintaining the Database:** It is easier to create the database based on ongoing research projects, but sometimes it becomes very difficult to maintain and update the database for future use.

6.2 Security and Confidentiality of Participating Utilities Database

Utility wastewater infrastructure data/information is fully protected by a Memorandum of Understanding and at multiple points in our Cyber-Infrastructure System, as described in the following steps below.

1. **Data is initially uploaded to the secure FTP server, web.cgit.vt.edu**

    a. Only the participating cities are provided with login credentials.

b. The data only resides on this server as a staging area, until it can be uploaded to its permanent storage location, after which time it is deleted from the FTP server.

2. **The data permanently resides on an Oracle Real Applications Clusters server at the SDSC.**

   a. The data is owned by a single Oracle schema, and the only other user accounts that exist in this database instance are assigned to approve researchers at Virginia Tech.

   b. The owners of these accounts must request access to each individual GIS feature layer.

   c. Access is only granted on a "need to know" basis, and user will receive read-only access.

   d. All IP addresses are blocked at the firewall at SDSC, with one exception: the ArcSDE middleware server we use as a gateway to access the data from ArcGIS.

3. **The ArcSDE server functions as our gateway to the data.**

   a. ArcSDE adheres to Oracle's authentication and authorization controls; any access that is prohibited in the database itself is likewise prohibited in ArcSDE.

   b. All IP addresses are blocked at the firewall on the ArcSDE server, with the exception of valid Virginia Tech IP addresses.

4. **The Fiber-Optic Communication Link between VT and SDSC.**

   a. The fiber-optic communications link between Virginia Tech and the San Diego Supercomputing Center is a dedicated point-to-point pathway.

   b. It is NOT part of the public Internet; rather, it is a link in the National Lambda Rail research network that is only used by major universities and institutions.

5. **San Diego Super Computer Center.**

   a. The primary goal of computer and network security activities at the SDSC is to protect the confidentiality, integrity, and availability of computing resources. As an Organized Research Unit at University of California, San Diego (UCSD), SDSC Security Policies are consistent with UC and UCSD policies. The SDSC Security Manager and authorized designees have the authority to take actions necessary to prevent, detect, and respond to security policy violations involving SDSC computing resources. These actions will be consistent with SDSC, UCSD, and UC policies, and applicable laws.

   b. The Data Intensive Computing Environments group at the SDSC implemented software to make use of a variety of authentication/data-security systems in the SDSC Storage Resource Broker and other distributed applications. These include the SSL(X.509 Certificate)-based Grid Security Infrastructure developed by Globus with assistance from National Center for Supercomputing Applications and others, Kerberos, Distributed Computing Environment Security, and SDSC Encryption/Authentication system.

CHAPTER 7.0

UTILITY CASE STUDIES

Chapter Highlights

- **Case Studies:** The following five utilities case studies are presented in this chapter:

    1. City of Atlanta, Atlanta, Georgia
    2. Pittsburgh Water and Sewer Authority, Pittsburgh, Pennsylvania
    3. Seattle Public Utility, Seattle, Washington
    4. Orange County Sanitation District, Fountain Valley, California
    5. VPI Sanitation Authority, Blacksburg, Virginia

To assist in research development and obtain crucial feedback about this current research, the research team visited various utilities and held net-meetings with utilities. Two utility field trips took place: Atlanta and Pittsburgh. On these trips the research met with city officials to discuss their progress toward the asset management of their wastewater collection systems. All city officials were very helpful and contributed to the wastewater infrastructure research. The cities realize the more help and input they provide to the research project, the more their utilities will benefit. Additional utility field trips will be scheduled in the future.

The research team held online meetings with participating utilities in order for the cities to present and depict their related wastewater pipe data in detail. It was shown that the cities had the same idea of compiling the data through GIS; however, many of the parameters are not within one system and must be derived indirectly. In past years, the utilities' pipe infrastructure data was kept in other forms besides an electronic database. Many utilities are in the process of transferring their data from paper formats and other sources into their GIS electronic databases. The research team prepared five case studies that give an overview of each city's wastewater system, utilities data collection, and utilities' GIS capabilities.

7.1 Case Study 1: Atlanta, Georgia

There are approximately 40,000 manholes and 1,900 miles of sewer main and laterals within right-of-way and easements in the City's wastewater collection and transmission systems. The combined sewers are estimated to be 85% of the system and the rest are separate sanitary sewers. The cities of Hapeville, College Park and East Point, and DeKalb, Clayton, and Fulton counties are six other entities that have wastewater treatment contracts with the city. The City of Atlanta generates 55% of sewage flows and the wholesale agencies have 45%. The total population served from the system is 1.6 million [89]. In 1999, Atlanta entered into a First

Amended Consent Decree with the U.S. EPA. The decree requires Atlanta to implement many of the programs associated with U.S. EPA's widely discussed wastewater collection system management initiative for capacity, management, operation and maintenance, known as CMOM. The City of Atlanta wastewater system is shown in Figure 7-1.

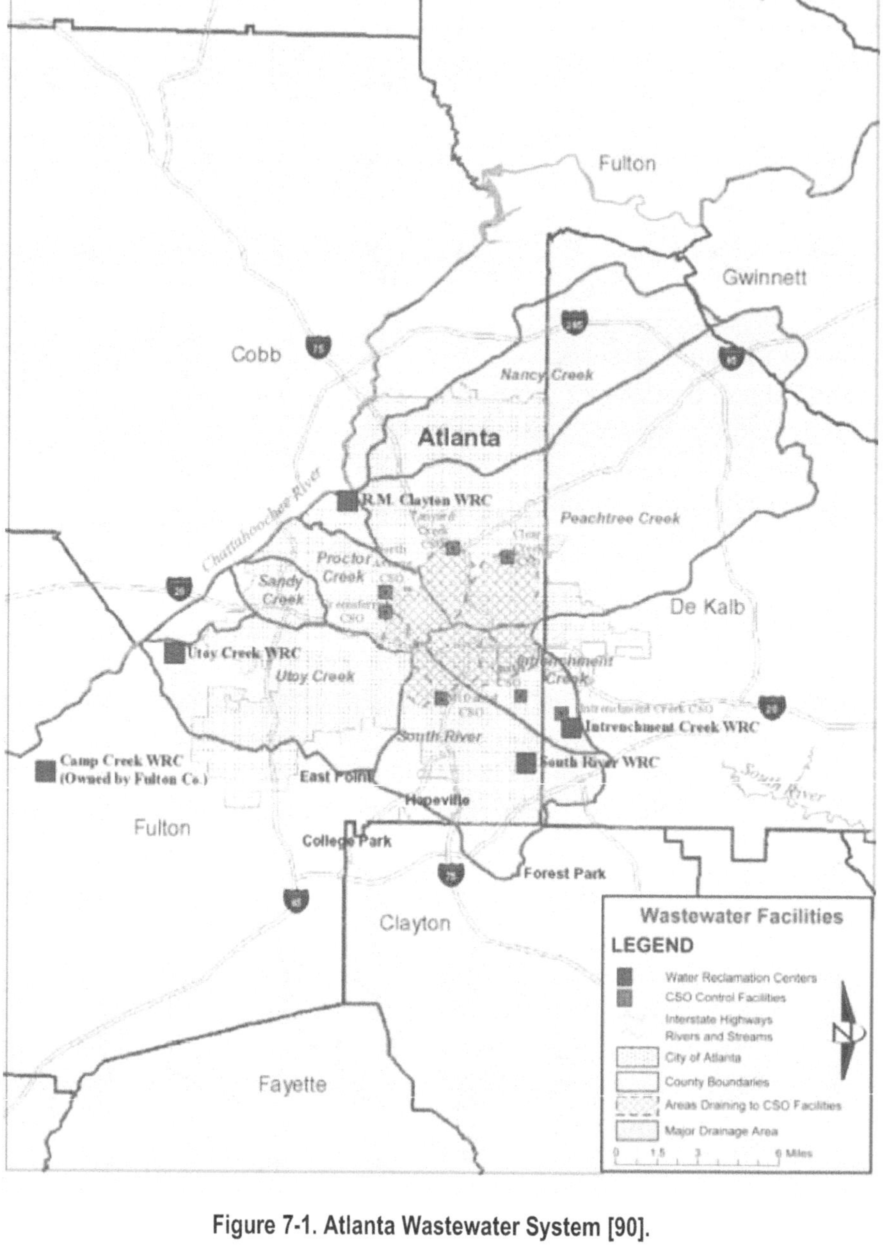

Figure 7-1. Atlanta Wastewater System [90].

The Atlanta Public Utilities Board (APUB) provided shapefiles, geo-database tables and work management tables stored in Hansen in a DVD, which is mailed to the research group at Virginia Tech. Altogether there were 14 shapefiles, 29 geo-database tables and approximately 1,200 Hansen© file tables. The shapefiles consists mainly of the main pipelines, laterals, sewer facilities consisting of manholes, rehab work related files along with transportation, soils, catch basins and city limits. A brief description of the main files given by APUB is given below.

- **Sewer Main**

The sewer main shapefile consists of all the sewer mains in Atlanta city limits that are spatially mapped in the projected coordinate system North American Datum (NAD)_1983_StatePlane_Georgia_West. The sewer main attribute table consists of various pipe parameters such as pipe ID, pipe mailing address (as in the street name, street number, pipe shape, diameter, material, length, and elevation of the upstream and downstream nodes. Also, the slope of the pipe was calculated and added as an attribute. The surface cover and geographical location of each pipe is given along with the sewer shed and basin references. Although a few columns were left blank, enough information was collected for around 40,000 pipes.

- **Sewer Facility**

The sewer facility file consists of manholes in Atlanta city limits that are spatially mapped in the projected coordinate system NAD_1983_StatePlane_Georgia_West. The shapefile consists of various manhole parameters such as facility ID, facility mailing address (as in the street name, street no), topographical location, depth of the manhole, diameter and the geographical coordinates. Some inner details like cover type, cover diameter, wall type are also given along with sewer shed and basin references. Although a few columns were left blank, enough information was collected for around 37,000 manholes.

- **SSES Tables**

The Sewer System Evaluation Survey (SSES) tables are geo-database tables that can be used to store general information that does not need to be spatially mapped. The SSES tables given by APUB have all the codes used by the utility officials. In general, location type, manhole type, street type codes are specified. Adding to these, pipe condition data, defects, type of leaks, inspection and inventory data are also stored. All these tables, however, contain a lot of null valued columns that might require some metadata to explain the purpose of the attributes.

- **Hansen© Tables**

The Hansen© tables given by APUB were numerous (around 1,200) most of which are empty. Hansen© is a work order management system used by the Atlanta utility officials. The tables mainly consist of the work orders issued by the officials. Some tables also consist of codes and other information that cannot be interpreted by the research team. In this regard, the APUB is contacted for further detailed information.

Table 7-1 presents the existing data parameters the research team received from APUB. Table 7-2 shows the data that can be derived or downloaded from existing sources.

Table 7-1. Existing Data Received from City of Atlanta (COA).

Parameter			
Physical/Structural		**Operational/Functional**	
1	*Node Identification Number	11	*Pipe Renewal Record
2	*Pipe Material	12	*Pipe Failure Record
3	*Pipe Diameter	13	*Infiltration/Inflow
4	*Pipe Age	14	*Exfiltration
5	*Pipe Location	15	*Blockage/Stoppage
6	*Pipe Shape	16	*Sediments
7	*Function of the Pipe	17	*Inspection Record
8	Pipe Lateral		
9	Lateral Connections		
10	Pumping Station and WWTP		

Note: * Essential Parameter for Gold Standard

Table 7-2. Derived or Downloadable Data for City of Atlanta (COA).

Parameter			
Physical/Structural		**Environmental** (Cont.)	
1	*Pipe Depth	18	*Extreme Events
2	*Node Length	19	*Soil Corrosivity
3	*Pipe Slope	20	*Soil Moisture Content
4	*Design Life	21	*Soil Type
5	*Design Strength	22	Runoff Rate
6	Pipe Quality	23	Soil pH
7	Pipe Vintage	24	Non-Uniform Soil
8	Dissimilar Materials	25	Non-Uniform Slope
9	Pipe Installation	26	Unstable Slope
Environmental		27	Seismic Activity
10	*Stray Currents	28	Catchment Area(sewershed)
11	*Groundwater Table	29	Average Closeness to Trees
12	*Ground Cover	30	Tidal Influences
13	*Loading Condition (dead load)	31	Pipe Connections
14	*Loading Condition (live load)	**Others**	
15	*Rainfall/Precipitation	32	Density of Connections
16	*Climate - Temperature	33	*Overall Pipe Condition
17	*Topography		

Note: * Essential Parameter for Gold Standard

Table 7-3 presents the needed data that research team will ask the utility to provide or search for from other sources.

Table 7-3. Missing Gold Standard Data for City of Atlanta (COA).

Parameter			
Physical/Structural		**Environmental**	
1	*Pipe Wall Thickness	26	*Soil Resistivity
2	*Pipe Joint Type	27	*Redox Potential
3	*Pipe Bedding	28	Soil Disturbance
4	*Trench Backfill	29	Frost Penetration
5	*Construction Specification	30	Soil Chloride
6	*Pipe Lining	31	Soil Sulfate
7	*Manhole Condition	32	Soil Sulfide
8	Pipe Section Length	**Financial**	
9	Pipe Manufacture	33	*Annual Capital Cost
10	Pipe Trench Width	34	Annual Maintenance Cost
11	Pipe External Coating	35	Annual Repair/Rehabilitation Cost
12	Pipe Cathodic Protection	36	Installed and Replacement Cost
13	Pipe Thrust Restraint	37	Annual Operational Cost
Operational/Functional		38	Annual Energy Cost
14	*Wastewater Quality	39	Depreciated Value
15	*Wastewater Pressure (Force Main)	40	Benefit/Cost
16	*Pipe Hydraulics	**Others**	
17	*Pipe Surcharging	41	*Customer Complaint
18	*Operational & Maintenance Practices	42	*Chemical
19	*Flow Velocity	43	*FOG
20	Sewer Odors	44	Resident Population Served
21	Sewer Flooding	45	Failing Utilities
22	Sewer Overflow (SSO/CSO)	46	Consequence/Risk
23	Backup Flooding	47	Third-party Damage
24	Leakage Allowance		
25	Interruptions		

Note: * Essential Parameter for Gold Standard

Condition Assessment and Rehabilitation Practice

The City of Atlanta will be inspecting every foot of the pipe in the system during condition assessment program, SSES. The goal of the SSES is to build a complete inventory of the wastewater system, to indentify sources of I&I, and perform a condition assessment. Atlanta's sewers have been categorized into six groups and the criteria for the initial prioritization of the Sewer Groups took account of 10 items: 1) frequency of overflows, 2) severity of rainfall-dependent /inflow, 3) risk to surface waters (creeks), 4) impact failure, 5) status of ongoing rehabilitation or renewal, 6) available capacity in sewers, 7) judgment of sewer operation and maintenance division, 8) relative impact of rainfall-dependent infiltration/inflow from jurisdiction outside the city's control, 9) proposed development intensity, and 10) location of sewer within the combined system. Six sewer groups will be studied over the 14-year compliance period of the first amended consent decree. With the findings from each SSES study an appropriate prioritized list is scheduled of sewers, manholes and laterals identified for remedial action [89].

- **Sewer Pipe Condition Assessment Practice**

In this program, sewer inspections are used to identify defects and sources of I&I including, CCTV, smoke test, dye test, flow monitoring, rain monitoring, building service connection location/inspection, and flow isolation. Defects are rated based on Atlanta's in-house defects rating, called internal condition grade (ICG) shown in Table 7-4. ICG gives a rating of 1 through 5 for each pipe section.

Table 7-4. Internal Condition Grade (ICG).

ICG	Typical Defect Description
5	Pipe or brick sewer already collapsed; or,
	Deformation exceeds 10% and pipe is broken; or,
	Extensive areas of material missing; or
	Deformation exceed 10% with fracture(s); or,
	Extreme loss of concrete/mortar
4	Pipe or brick sewer is broken; or,
	Deformation <10% with fracture(s); or,
	Multiple fractures; or,
	Serious loss of gradient; or,
	Severe concrete corrosion; or, bricks displaced
3	Pipe or brick sewer is fractured with deformation between 5% and 10%; or,
	Longitudinal cracking or multiple cracking; or,
	Severe joint defects; or,
	Badly made connections; or,
	Moderate concrete corrosion; or,
	Bricks displaced
2	Light corrosion; or circumferential cracks; or moderate joint defects
1	No corrosion or structural defects

- **Sewer Pipe Repair/Rehabilitation/Replacement Practice**

GIS, CCTV videos, smoke testing photos, manhole inspection reports, spill reports, hydraulic modeling, and rehabilitation cost are stored in SSES Raw database as illustrated in Figure 7-2. This information from SSES database as well as defect rating is used to support the decision making. The designer determines the main rehabilitation approach according to the guidelines and identifies the approach using Rehabilitation Selection Tool. The main rehabilitation selection is based on the presence and frequency of a group of essential structural defects such as breaks, holes, multiple fractures, deformations, and large joint displacements. Extensive service defects such as multiple root masses and root taps may also warrant rehabilitation action.

Using Rehabilitation Selection Tool, engineers categorize projects into Capacity Relief Projects; Rehabilitation; Capacity assessment, maintenance, operation, and management (CMOM) and Deferral; No Rehabilitation. The Capacity Relief Projects category is specified for all trunks and large sewers (greater than 24-inch diameter). Rehabilitation is the main category and is specified for trenchless rehabilitation as well as traditional open cut replacement methods. This category also includes root removal, cleaning and manhole rehabilitation and replacement. CMOM and Deferral is for sewers for which action is deferred and for sewers that need to be monitored for a period of time before a final rehabilitation decision is to be made. In cases where only minor defects or no defects are present, the sewers are placed in the No Rehabilitation category [89].

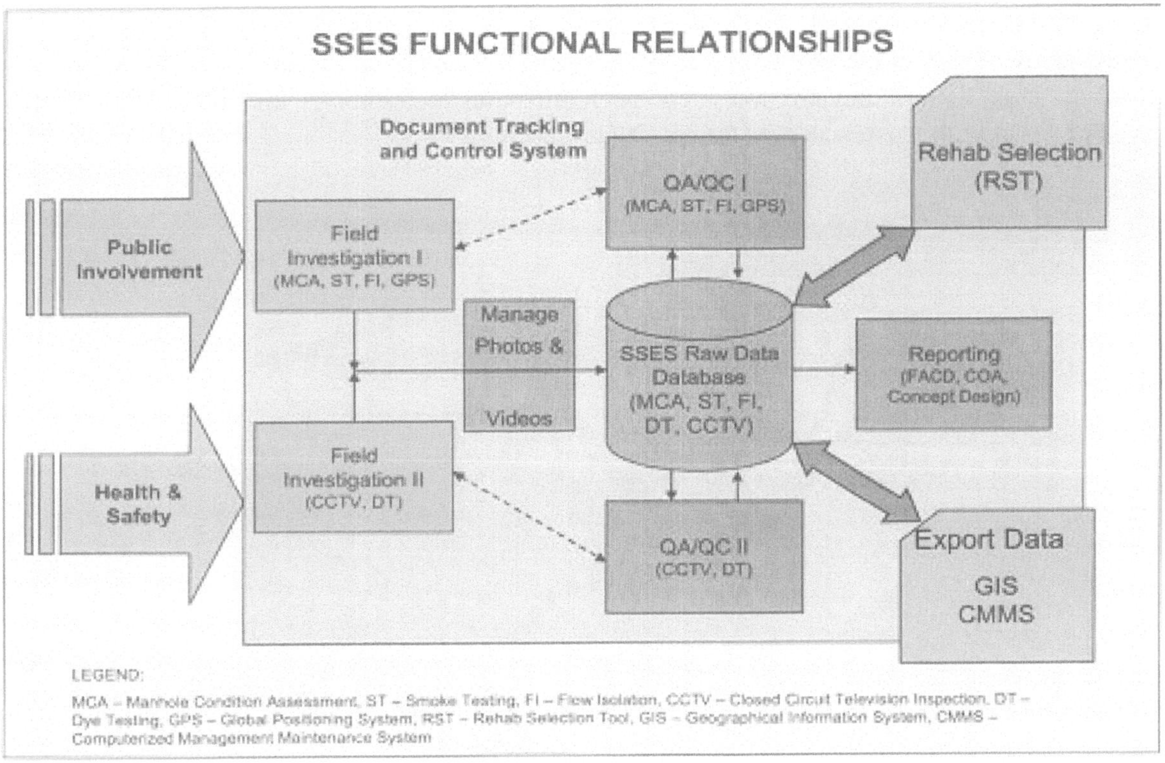

Figure 7-2. Atlanta's SSES Program [90].

7.2 Case Study 2: Pittsburgh, Pennsylvania

In 1995, the City of Pittsburgh's Water Department became a part of Pittsburgh Water and Sewer Authority (PWSA). In 1999, PWSA became responsible for operating and maintaining the entire City of Pittsburgh sewer system. Founded in 1984, PWSA serves approximately 250,000 consumers throughout the City of Pittsburgh [91].

Recently PWSA proposed the combined sewer overflow (CSO) program, which aims to identify cost-effective CSO control alternatives that, when fully implemented, protect water quality [92]. The development of this program required monitoring and sampling plans that proposed data collection and characterization activities to be a CSO Long-term Control Plan (LTCP). The CSO LTCP is the characterization of the combined sewer system operation and the assessment of CSO impacts on river and stream water quality during wet weather events. This monitoring and sampling plan presents the proposed data collection and characterization activities to be undertaken.

PWSA and their consultant have given shapefiles in a DVD, which is mailed to the researchers. Altogether there were eight shapefiles, mainly of the main pipelines, manholes, junctions, and diameter changes. A brief description of the main files given by PWSA is described below. One unique characteristic of these shapefiles is that the same attributes are listed for all, even though they are unrelated. Such attributes are either given a null value or left empty. The Pittsburgh wastewater system is shown in Figure 7-3.

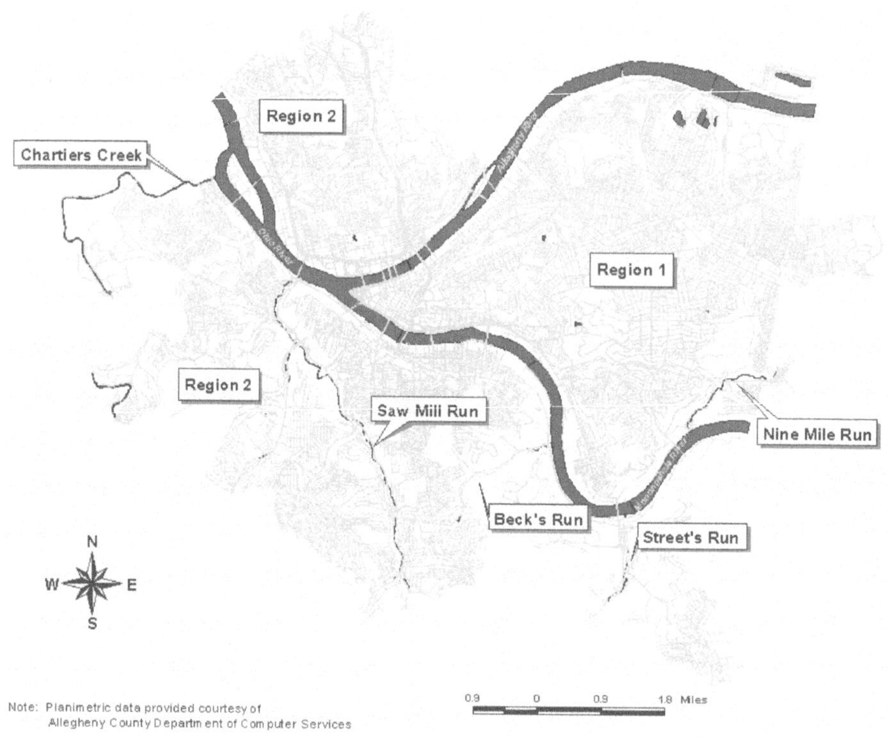

Figure 7-3. Pittsburgh Wastewater System [92].

- **Mainline**

The sewer main shapefile consists of all the sewer mains in Pittsburgh city limits that are spatially mapped in an undefined projected coordinate system. The sewer main attribute table consists of various pipe parameters such as pipe ID, pipe shape, diameter, material, length, installation year, and elevation of the upstream and downstream nodes. Although a few columns were left blank, enough information was collected for around 44,000 pipes.

- **Manhole**

This consists of all the manholes in Pittsburgh city limits that are spatially mapped in an undefined projected coordinate system. The shapefile consists of various manhole parameters such as facility ID, depth of the manhole, diameter, and the geographical coordinates. Although a few columns were left blank, enough information was collected for around 30,000 manholes.

- **Diameter Change**

This is a unique collection of data from other utilities. All the diameter change points are noted and spatially mapped in an undefined coordinate system. The junction points are given a node ID and the types of reducers used are listed. Table 7-5 presents the existing data parameters the research team received from PWSA.

Table 7-5. Existing Data Received from Pittsburgh Water and Sewer Authority (PWSA).

Parameter			
Physical/Structural			
1	*Node Identification number	6	*Pipe Shape
2	*Pipe Material	7	*Function of the Pipe
3	*Pipe Diameter	8	Pipe Lateral
4	*Pipe Age	9	Lateral Connections
5	*Pipe Location	10	Pumping Station and WWTP

Note: * Essential Parameter for Gold Standard.

Table 7-6 shows the data that can be derived or downloaded from existing sources.

Table 7-6. Derived or Downloadable Data for Pittsburgh Water and Sewer Authority (PWSA).

Parameter			
Physical/Structural		**Environmental** (Cont.)	
1	*Pipe Depth	17	*Topography
2	*Node Length	18	*Extreme Events
3	*Pipe Slope	19	*Soil Corrosivity
4	*Design Life	20	*Soil Moisture Content
5	*Design Strength	21	*Soil Type
6	Pipe Quality	22	Runoff Rate
7	Pipe Vintage	23	Soil pH
8	Dissimilar Materials	24	Non-Uniform Soil
9	Pipe Installation	25	Non-Uniform Slope
Environmental		26	Unstable Slope
10	*Stray Currents	27	Seismic Activity
11	*Groundwater Table	28	Catchment Area (sewershed)
12	*Ground Cover	29	Average Closeness to Trees
13	*Loading Condition (dead load)	30	Tidal Influences
14	*Loading Condition (live load)	31	Pipe Connections
15	*Rainfall/Precipitation	**Others**	
16	*Climate - Temperature	32	Density of Connections

Note: * Essential Parameter for Gold Standard

Table 7-7 presents the needed data the research team will ask the utility to provide or search for from other sources.

Table 7-7. Missing Gold Standard Data for Pittsburgh Water and Sewer Authority (PWSA).

Parameter			
Physical/Structural		**Environmental**	
1	*Pipe Wall Thickness	33	*Soil Resistivity
2	*Pipe Joint Type	34	*Redox Potential
3	*Pipe Bedding	35	Soil Disturbance
4	*Trench Backfill	36	Frost Penetration
5	*Construction Specification	37	Soil Chloride
6	*Pipe Lining	38	Soil Sulfate
7	*Manhole Condition	39	Soil Sulfide
8	Pipe Section Length	**Financial**	
9	Pipe Manufacture	40	*Annual Capital Cost
10	Pipe Trench Width	41	Annual Maintenance Cost
11	Pipe External Coating	42	Annual Repair/Rehabilitation Cost
12	Pipe Cathodic Protection	43	Installed and Replacement Cost
13	Pipe Thrust Restraint	44	Annual Operational Cost
Operational/Functional		45	Annual Energy Cost
14	*Wastewater Quality	46	Depreciated Value
15	*Wastewater Pressure (Force Main)	47	Benefit/Cost
16	*Pipe Hydraulics	**Others**	
17	*Pipe Surcharging	48	*Customer Complaint
18	*Operational & Maintenance Practices	49	*Chemical
19	*Flow Velocity	50	*FOG
20	*Pipe Renewal Record	51	*Overall Pipe Condition
21	*Pipe Failure Record	52	Resident Population Served
22	*Infiltration/Inflow	53	Failing Utilities
23	*Exfiltration	54	Consequence/Risk
24	*Blockage/stoppage	55	Third-party Damage
25	*Sediments		
26	*Inspection Record		
27	Sewer Odors		
28	Sewer Flooding		
29	Sewer Overflow (SSO/CSO)		
30	Backup Flooding		
31	Leakage Allowance		
32	Interruptions		

Note: * Essential Parameter for Gold Standard

Condition Assessment and Rehabilitation Practice

The PWSA team developed a CCTV program that schedules the inspection of "critical sewers." A review of the sewer system mapping was performed to highlight the sewers that will be modeled including the preliminary flow monitoring locations. Approximately 80 miles of sewer main was designated for televising. The individual sewer segments within each of the watersheds have been preliminarily identified for CCTV investigation. The development of the sewer segment list will be a working document. As additional information is gathered from the sewer investigation field activities and through discussions with PWSA staff, the sewer segment selections will be amended. If sections of the main truck line sewers are scheduled for rehabilitation or replacement, or if recent televising work was completed, those sections may not need to be televised during this project. Video data tapes, digital video data on CD-ROM, and television logs will be created for each of the televised sewer segments.

- **Sewer Pipe Condition Assessment Practice**

CCTV still frame video images and video clips in a standard video image format will be transferred to the GIS, for use in the InfoNet Commercial Software Database. All staff responsible for rating and evaluating the sewer system conditions shall be NASSCO certified.

The selection of the CCTV segments will be based on:

- The field inspection program findings that suggest an operational and/or maintenance issue exists in the system that may impact the hydraulics,
- Record drawings or other information necessary to reconcile the understanding of the sewer network does not exist or cannot be obtained
- Inspection of the pipe is needed to satisfy requirements of the forthcoming CSO consent order.

The CCTV condition assessment program will target areas identified as "critical sewers." These areas will be included in model development, as well as additional areas required to clarify system boundaries and connectivity.

- **Sewer Pipe Failure Prediction**

The primary objectives of the hydrologic/hydraulic model of the PWSA collection system are to:

- Characterize the hydraulic response of the sewer system during a variety of wet weather events and calibrate modeled versus metered flows,
- Facilitate the analysis of CSO control alternatives, and
- Predict pollutant loadings to the receiving waters.

Once the model is developed and calibrated, it will be used as a predictive tool to support CSO abatement planning efforts.

7.3 Case Study 3: Seattle, Washington

Seattle Public Utilities (SPU) owns and operates the wastewater collection system for the city of Seattle, which has 1,491 miles of combined sewer and sanitary pipelines. The wastewater from the city is treated at King County sewage treatment plants. The city's wastewater system services 570,000 people [93]. SPU's management system includes, a 20-year comprehensive plan providing long-term direction setting, a three-year strategic plan setting the objectives and targets in line with the comprehensive plan; and specific management systems supporting the objectives and targets of the strategic plan [93]. The system plan, set in 2006, focused on describing all of SPU's existing wastewater policies and identified areas where additional policies need to be developed; identifying levels of customer service to minimize sewer backups, street flooding in combined sewer areas, control of combined sewer overflows, and emergency response; presenting strategies and an implementation plan to meet the established customer service levels; establishing a financial program to fund the programs and activities in the plan [94].

SPU provided the researchers with shapefiles, geoguide, CCTV, and Excel files. Altogether there were 14 shapefiles along with a geoguide. The geoguide served as a metadata file that describes all the attributes mentioned in the shapefiles, including main pipelines, laterals, manholes, catch basins and drainage basins, along with transportation and terrain. A brief description of the main files given by SPU is given below.

- **Mainlines**

This shapefile consists of all the sewer mains in Seattle city limits that are spatially mapped in the projected coordinate system NAD_1983_StatePlane_Washington_South. The sewer main attribute table consists of various pipe parameters such as pipe ID, pipe shape, diameter, installation year, material, length, elevation of the upstream and downstream nodes Although a few columns were left blank, enough information was collected for around 55,000 pipes.

- **Manholes**

This shapefile consists of manholes within Seattle city limits that are spatially mapped in the projected coordinate system NAD_1983_StatePlane_Washington_South. The shapefile consists of various manhole parameters such as facility ID, topographical location, depth of the manhole, diameter, and the geographical coordinates. Some inner details like case type, probable flow are also given. Although a few columns were left blank, enough information was collected for around 55,000 manholes.

- **Drainage Basins**

This shapefile is a collection of parameters of the drainage basins in Seattle city limits like area, perimeter and outfall numbers. Nearest waterbodies are also listed along with the catalog IDs of the basins. Altogether there are 159 drainage basins.

Table 7-8 presents the existing data parameters Virginia Tech research team received from SPU.

Table 7-8. Existing Data Received from Seattle Public Utility (SPU).

Parameter			
Physical/Structural		**Operational/Functional**	
1	*Node Identification number	11	*Pipe Failure Record
2	*Pipe Material	12	*Infiltration/Inflow
3	*Pipe Diameter	13	*Exfiltration
4	*Pipe Age	14	*Blockage/stoppage
5	*Pipe Location	15	*Sediments
6	*Pipe Shape	16	*Inspection Record
7	*Function of the Pipe		
8	Pipe Lateral		
9	Lateral Connections		
10	Pumping Station and WWTP		

Note: * Essential Parameter for Gold Standard

Table 7-9 shows the data that can be derived or downloaded from existing sources.

Table 7-9. Derived or Downloadable Data for Seattle Public Utility (SPU).

Parameter			
Physical/Structural		**Environmental** (Cont.)	
1	*Pipe Depth	18	*Extreme Events
2	*Node Length	19	*Soil Corrosivity
3	*Pipe Slope	20	*Soil Moisture Content
4	*Design Life	21	*Soil Type
5	*Design Strength	22	Runoff Rate
6	Pipe Quality	23	Soil pH
7	Pipe Vintage	24	Non-Uniform Soil
8	Dissimilar Materials	25	Non-Uniform Slope
9	Pipe Installation	26	Unstable Slope
Environmental		27	Seismic Activity
10	*Stray Currents	28	Catchment Area (Sewer-shed)
11	*Groundwater Table	29	Average Closeness to Trees
12	*Ground Cover	30	Tidal Influences
13	*Loading Condition (dead load)	31	Pipe Connections
14	*Loading Condition (live load)	**Others**	
15	*Rainfall/Precipitation	32	Density of Connections
16	*Climate - Temperature	33	*Overall Pipe Condition
17	*Topography		

Note: * Essential Parameter for Gold Standard

Table 7-10 presents the data that research team will request from the utility or from other sources.

Table 7-10. Missing Gold Standard Data for Seattle Public Utility (SPU).

Parameter			
Physical/Structural		**Environmental**	
1	*Pipe Wall Thickness	27	*Soil Resistivity
2	*Pipe Joint Type	28	*Redox Potential
3	*Pipe Bedding	29	Soil Disturbance
4	*Trench Backfill	30	Frost Penetration
5	*Construction Specification	31	Soil Chloride
6	*Pipe Lining	32	Soil Sulfate
7	*Manhole Condition	33	Soil Sulfide
8	Pipe Section Length	**Financial**	
9	Pipe Manufacture	34	*Annual Capital Cost
10	Pipe Trench Width	35	Annual Maintenance Cost
11	Pipe External Coating	36	Annual Repair/Rehabilitation Cost
12	Pipe Cathodic Protection	37	Installed and Replacement Cost
13	Pipe Thrust Restraint	38	Annual Operational Cost
Operational/Functional		39	Annual Energy Cost
14	*Wastewater Quality	40	Depreciated Value
15	*Wastewater Pressure (Force Main)	41	Benefit/Cost
16	*Pipe Hydraulics	**Others**	
17	*Pipe Surcharging	42	*Customer Complaint
18	*Operational & Maintenance Practices	43	*Chemical
19	*Flow Velocity	44	*FOG
20	*Pipe Renewal Record	45	Resident Population Served
21	Sewer Odors	46	Failing Utilities
22	Sewer Flooding	47	Consequence/Risk
23	Sewer Overflow (SSO/CSO)	48	Third-party Damage
24	Backup Flooding		
25	Leakage Allowance		
26	Interruptions		

Note: * Essential Parameter for Gold Standard

Condition Assessment and Rehabilitation Practice

SPU began the implementation of an asset management program about five years ago with the initial goal to inventory the existing pipe infrastructure and develop a modeling methodology that would provide the foundation for a successful risk-based replacement/rehabilitation strategy. The immediate goal was to minimize the risk that infrastructure failures were likely bringing to the utility [95]. At present, SPU condition assessment decisions are based on the Sewer Pipe Risk Model, an in-house model. Fundamentally, the Sewer Pipe Risk Model calculates the risk cost of failure for each individual pipe by multiplying the estimated consequence of failure by the estimated likelihood of failure.

♦ Sewer Pipe Condition Assessment Practice

SPU has used PACP coding on wastewater pipes since 2005. Before using PACP coding, a customized coding system developed in-house had been used since the late 1960s. SPU primarily uses conventional CCTV inspection methods by using a video camera mounted on a remote-controlled tractor unit and uses a zoom camera on pipes to calibrate various types of maintenance frequencies, but not to determine structural condition. The pipe inspection was completed and the GIS was updated on a regular basis. Inspections were prioritized for the last five years using Sewer Pipe Risk Model. The risk-based strategy identified pipes that are critical and need to be inspected regularly as shown in Figure 7-4.

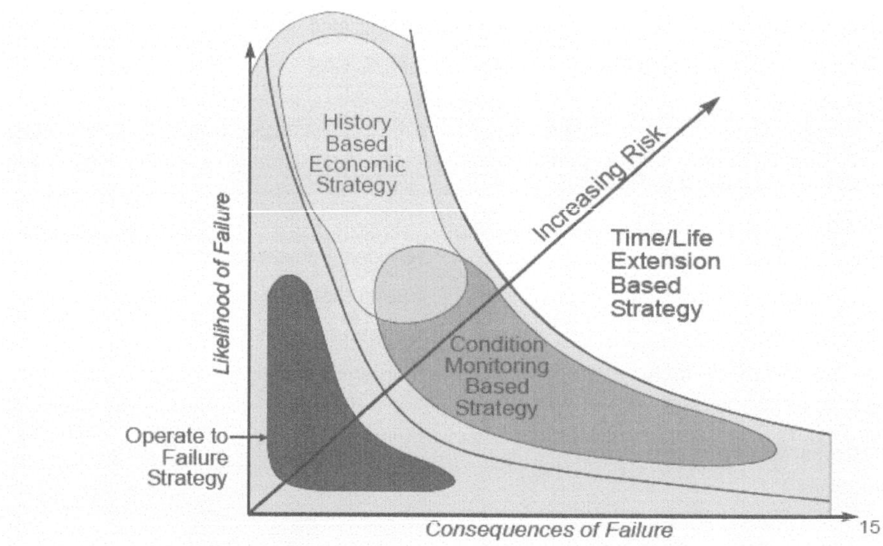

Figure 7-4. Strategy for Management of Assets [96].

♦ Sewer Pipe Repair/Rehabilitation/Replacement Practice

Rehabilitation is currently prioritized by using an expert system although SPU is close to implementing an automated sewer pipe rehab decision model developed by an academic/consultant. Currently though CCTV tapes are PACP-coded and a manual decision is made on whether the pipe should have no action, a single point repair, a series of point repairs, or a re-line. Full pipe replacement is almost never done because SPU found it not to be cost effective.

- ♦ **Sewer Pipe Failure Prediction**

SPU uses predictive failure curves to aid in calculating the risk cost of its sewer pipes. The time-based predictive failure curves use a normalized Weibull-based distribution. It applies a failure curve to each pipe in the system based on age and material type. Recently, an analysis was undertaken to verify the accuracy of SPU's existing curves and to possibly replace them with an expected point failure distribution more precisely based on Seattle's actual sewer pipe failure history. Figure 7-5 shows likelihood of pipe failure calculated by normalized Weibull-based distribution.

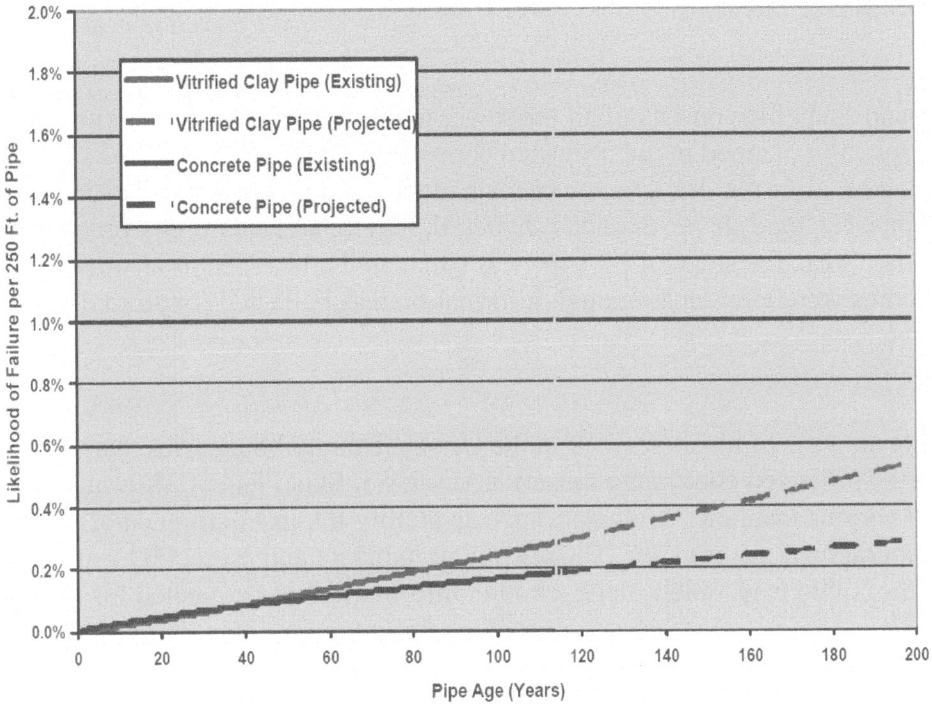

Figure 7-5. Likelihood of Pipe Failure [95].

7.4 Case Study 4: Orange County, California

The Orange County Sanitation District (OCSD), formed in 1946, collects, treats, disposes of, and reclaims the wastewater generated by 2.5 million people in central and northwestern Orange County [97]. OCSD includes nine former revenue areas joined into a single service district, forming the third largest wastewater agency in the western United States [93]. OCSD operates two treatment plants and maintains 580 mile of wastewater pipes. 250 million gallons of wastewater flow through daily through 16 pumping stations. One treatment plant is located in Fountain Valley and the other in Huntington Beach.

For the purpose of recycling water, OCSD partnered with Orange County Water District on the Green Acre Project (GAP) and Ground Water Replenishment System. GAP reclaims 10 mgd treated wastewater via microfiltration and reverse osmosis [97]. This reclaimed water is used for landscape irrigation. For the Ground Water Replenishment System, 70 mgd treated wastewater is reclaimed, after being purified through microfiltration, reverse osmosis, and ultraviolet disinfection to levels that exceed drinking water standards [93]. This water is used for injection into the groundwater to act as a seawater intrusion barrier and for replenishment of the local drinking water aquifer. In addition, OCSD undertook management systems initiatives in two main areas: optimized asset management and the National Biosolids Partnership Environmental Management System. OCSD also engaged in strategic planning activities and created the unifying strategies [93].

The OCSD provided the researchers with a DVD of 13 shapefiles, including the main pipelines, manholes, land use along with soils and city limits. A brief description of the main files given by OCSD is given below. The sewer system in the city also has some forced main pipes along with fittings and valves.

- **Mainlines**

The sewer main shapefile consists of all the sewer mains within Orange County boundaries that are spatially mapped in the projected coordinate system NAD_1983_StatePlane_California_VI. The sewer main attribute table consists of various pipe parameters such as pipe ID, pipe shape, diameter, material, length, and elevation of the upstream and downstream nodes. Also, the slope of the pipe was calculated and added as an attribute. Although a few columns were left blank, enough information was collected for around 9000 pipes.

- **Manholes**

This shapefile consists of manholes within the Orange County boundaries that are spatially mapped in the projected coordinate system NAD_1983_StatePlane_California_VI. The shapefile consists of various manhole parameters such as facility ID, installation date, depth, diameter, and the geographical coordinates. The document numbers and project ID's are also noted. Although a few columns were left blank, enough information was collected for around 8000 manholes.

Table 7-11 presents the existing data parameters the research team received from OCSD.

Table 7-11. Existing Data Received from Orange County Sanitation District (OCSD).

Parameter			
Physical/Structural			
1	*Node Identification number	6	*Pipe Shape
2	*Pipe Material	7	*Function of the Pipe
3	*Pipe Diameter	8	Pipe Lateral
4	*Pipe Age	9	Lateral Connections
5	*Pipe Location	10	Pumping Station and WWTP

Note: * Essential Parameter for Gold Standard

Table 7-12 shows the data that can be derived or downloaded from existing sources.

Table 7-12. Derived or Downloadable Data for Orange County Sanitation District (OCSD).

Parameter			
Physical/Structural		**Environmental** (Cont.)	
1	*Pipe Depth	17	*Topography
2	*Node Length	18	*Extreme Events
3	*Pipe Slope	19	*Soil Corrosivity
4	*Design Life	20	*Soil Moisture Content
5	*Design Strength	21	*Soil Type
6	Pipe Quality	22	Runoff Rate
7	Pipe Vintage	23	Soil pH
8	Dissimilar Materials	24	Non-Uniform Soil
9	Pipe Installation	25	Non-Uniform Slope
Environmental		26	Unstable Slope
10	*Stray Currents	27	Seismic Activity
11	*Groundwater Table	28	Catchment Area(sewershed)
12	*Ground Cover	29	Average Closeness to Trees
13	*Loading Condition (dead load)	30	Tidal Influences
14	*Loading Condition (live load)	31	Pipe Connections
15	*Rainfall/Precipitation	**Others**	
16	*Climate - Temperature	32	Density of Connections

Note: * Essential Parameter for Gold Standard

Table 7-13 presents the needed data that the research team will ask the utility to provide or search for from other sources.

Table 7-13. Missing Gold Standard Data for Orange County Sanitation District (OCSD).

Parameter			
Physical/Structural		**Environmental**	
1	*Pipe Wall Thickness	33	*Soil Resistivity
2	*Pipe Joint Type	34	*Redox Potential
3	*Pipe Bedding	35	Soil Disturbance
4	*Trench Backfill	36	Frost Penetration
5	*Construction Specification	37	Soil Chloride
6	*Pipe Lining	38	Soil Sulfate
7	*Manhole Condition	39	Soil Sulfide
8	Pipe Section Length	**Financial**	
9	Pipe Manufacture	40	*Annual Capital Cost
10	Pipe Trench Width	41	Annual Maintenance Cost
11	Pipe External Coating	42	Annual Repair/Rehabilitation Cost
12	Pipe Cathodic Protection	43	Installed and Replacement Cost
13	Pipe Thrust Restraint	44	Annual Operational Cost
Operational/Functional		45	Annual Energy Cost
14	*Wastewater Quality	46	Depreciated Value
15	*Wastewater Pressure (Force Main)	47	Benefit/Cost
16	*Pipe Hydraulics	**Others**	
17	*Pipe Surcharging	48	*Customer Complaint
18	*Operational & Maintenance Practices	49	*Chemical
19	*Flow Velocity	50	*FOG
20	*Pipe Renewal Record	51	*Overall Pipe Condition
21	*Pipe Failure Record	52	Resident Population Served
22	*Infiltration/Inflow	53	Failing Utilities
23	*Exfiltration	54	Consequence/Risk
24	*Blockage/stoppage	55	Third-party Damage
25	*Sediments		
26	*Inspection Record		
27	Sewer Odors		
28	Sewer Flooding		
29	Sewer Overflow (SSO/CSO)		
30	Backup Flooding		
31	Leakage Allowance		
32	Interruptions		

Note: * Essential Parameter for Gold Standard

- **Condition Assessment and Rehabilitation Practice**

OCSD adopted an asset management strategic plan and framework analysis in December 2002. It developed its first asset management plan in 2005 and a second version in 2006 with the goal to create, maintain, and rehabilitate the wastewater assets in the most cost effective (lowest life cycle cost) sustainable manner at the level of service required [98]. In regards to condition assessment, the basic functional processes of this asset management plan are to determine the current condition and performance, to determine the asset's likely failure modes and the probable time of failure. The failure modes include condition or structural failure, end of useful life, capacity, not meeting an established level of service, and no longer economic to own and operate.

- **Sewer Pipe Condition Assessment Practice**

CCTV and man entry in manholes are the techniques to inspect wastewater pipes. Also NASSCO standards are referenced. Inspections are based on a number of factors such as the following: age, H_2S levels, CCTV data, business risk exposure (BRE), depth of main, land use areas above pipes, and flow rate. Small diameter pipes are inspected randomly by CCTV based on historical understanding of the area. If spills have occurred, the line will be re-inspected. Routine inspections are performed for all manholes, and large diameter pipes are CCTV-inspected every 7 years. Small pipes with historical root intrusion problems have root cutting scheduled every 18 months.

OCSD has two separate database systems. Wallingford's software system is used for database management and is periodically updated. A GIS software system is updated when information is available. NASSCO codes are used with the GIS mapping to graphically represent pipe condition.

- **Sewer Pipe Repair/Rehabilitation/Replacement Practice**

Prioritized repair and rehab are based on structural integrity issues, life cycle costs, NASSCO codes (high levels), and age of pipes. Also, structural integrity compromise is considered, specifically the corrosion rate of pipe material. As an example, if more than 25% of rebar is exposed because of corrosion, and then replacement will occur for that section.

Business risk exposure (Figure 7-6) has been developed for the collection system, to prioritize asset condition assessments and cleaning. The outputs of this model are used to identify potential future capital improvement program (CIP) projects and are used with capacity modeling currently being undertaken. This was based H_2S concentration levels, age, depth, capacity, service area (urban vs. metropolitan), and NASSCO standards.

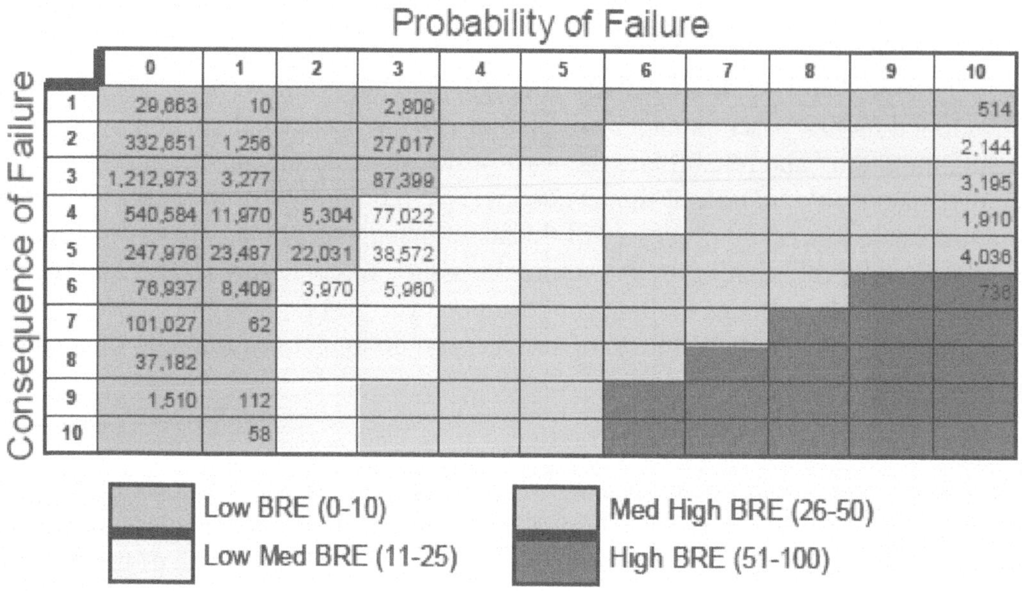

Figure 7-6. Business Risk Exposure by Length (ft).

An appropriate management strategy will be allocated based on business risk exposure. Examples of appropriate strategies are:

- Schedule regular condition assessments or monitoring
- CCTV – only clean pipes that represent the highest risk for both failure modes
- Do not CCTV – only assess pipes in a line that represent significantly different ratings
- Select a representative sample from each material type and age and apply results to other assets of a similar material and age
- Establish work order for future inspections in computerized maintenance management system (CMMS)
- Status quo (continue current strategy)
- Do nothing
- Renew (repair /rehabilitate /replace)
- Change maintenance /operations levels (e.g. Run to Failure)
- Upgrade (modify operational functionality – change performance requirements)
- Remove (not used /unnecessary)

♦ Sewer Pipe Failure Prediction

OCSD uses deterioration curves that are based on an age model and H_2S categories. The average age of the system is increasing. On the basis of the predicted maximum potential lives of the assets, the maximum life that gravity pipes can reach is 130 years old. Also, Pomeroy sulfide flux model calculations are used to help determine corrosion rate and NASSCO standards are referenced to help determine life expectancy.

7.5 Case Study 5: Blacksburg, Virginia

Blacksburg VPI Sanitation Authority Inc. and the Town of Blacksburg operate the wastewater system. Founded in 1962, VPI Sanitation Authority is a non-profit organization that is a collaborative effort between the Town of Blacksburg and Virginia Tech. The authority owns and operates one treatment plant at Stroubles Creek, and has about 15 miles of sewer main. There are 14 employees. All sanitary and industrial wastewater treated by the plant comes from three customers: the Town of Blacksburg, Virginia Tech, and part of Montgomery County, VA. Proportions of the wastewater collected from each customer are calculated based on their tap water usage. On average, 22%, 75%, and 3% of wastewater treated by the plant is from Virginia Tech, the Town of Blacksburg, and Montgomery County, respectively.

The Town of Blacksburg owns and operates 21 pumping stations and about 145 miles of sanitary sewer. The Town's system is divided into 17 sewer sheds. Each of these basins is further divided into sub-sewer sheds.

The Town of Blacksburg Utility provided two shapefiles and one geo-database tables to the researchers. The shapefiles consist of the main pipelines, manholes and a sewer shed geo-database table. A brief description of the main files is given below.

- **Mainlines**

The sewer main shapefile consists of all the sewer mains in Town of Blacksburg limits that are spatially mapped in the projected coordinate system NAD_1983_StatePlane_Virginia_South. The sewer main attribute table consists of various pipe parameters such as pipe ID, diameter, material, section length, joint type, street address, connected manhole, manhole condition, and pipe condition. Also, the length of the pipe was calculated and added as an attribute. Although a few columns were left blank, enough information was collected for about 4700 pipes. A screenshot of Blacksburg's GIS is shown in Figure 7-7.

Figure 7-7. Town of Blacksburg's GIS Data.

- **Manholes**

This shapefile consists of manholes in Town of Blacksburg limits that are spatially mapped in the projected coordinate system NAD_1983_StatePlane_Virginia_South. Except for the manhole condition, which presents in mainline shapefiles, there is a small amount of information in the manhole shapefile.

Table 7-14 presents the existing data parameters the research team received from Blacksburg Utility.

Table 7-14. Existing Data Received from Town of Blacksburg.

Parameter			
Physical/Structural			
1	*Node Identification number	6	*Pipe Joint Type
2	*Pipe Material	7	*Manhole Condition
3	*Pipe Diameter	8	*Overall Pipe Condition
4	*Pipe Location	9	Pipe Section Length
5	*Pipe Length		

Note: * Essential Parameter for Gold Standard

Table 7-15 shows the data that can be derived or downloaded from existing sources.

Table 7-15. Derived or Downloadable Data for Town of Blacksburg.

Parameter			
Physical/Structural		**Environmental (Cont.)**	
1	Dissimilar Materials	13	Unstable Slope
Environmental		14	Seismic Activity
2	*Stray Currents	15	Catchment Area (sewershed)
3	*Groundwater Table	16	Average Closeness to Trees
4	*Ground Cover	17	Tidal Influences
5	*Loading Condition (dead load)	18	Pipe Connections
6	*Loading Condition (live load)	19	*Soil Corrosivity
7	*Rainfall/Precipitation	20	*Soil Moisture Content
8	*Climate - Temperature	21	Runoff Rate
9	*Topography	22	Soil pH
10	*Extreme Events	23	*Soil Type
11	Non-Uniform Soil	**Others**	
12	Non-Uniform Slope	24	Density of Connections

Note: * Essential Parameter for Gold Standard

Table 7-16 presents the needed data that the research team will ask the utility to provide or search for from other sources.

Table 7-16. Missing Gold Standard Data for Town of Blacksburg.

Parameter			
Physical/Structural		**Operational/Functional** (Cont.)	
1	*Pipe Age	36	*Flow Velocity
2	*Pipe Depth	37	Sewer Odors
3	*Pipe Wall Thickness	38	Sewer Flooding
4	*Pipe Shape	39	Sewer Overflow (SSO/CSO)
5	*Function of Pipe	40	Backup Flooding
6	*Pipe Bedding	41	Leakage Allowance
7	*Trench Backfill	42	Interruptions
8	*Construction Specification	**Environmental**	
9	*Pipe Slope	43	*Soil Resistivity
10	*Pipe Design Life	44	*Redox Potential
11	*Pipe Design Strength	45	Soil Disturbance
12	*Pipe Lining	46	Frost Penetration
13	Pipe Quality	47	Soil Chloride
14	Pipe Vintage	48	Soil Sulfate
15	Pipe Lateral	49	Soil Sulfide
16	Pipe Installation	**Financial**	
17	Pipe Manufacture	50	*Annual Capital Cost
18	Pipe Trench Width	51	Annual Maintenance Cost
19	Pipe External Coating	52	Annual Repair/Rehabilitation Cost
20	Pipe Cathodic Protection	53	Installed and Replacement Cost
21	Pipe Thrust Restraint	54	Annual Operational Cost
22	Lateral Connections	55	Annual Energy Cost
23	Pumping Station and WWTP	56	Depreciated Value
Operational/Functional		57	Benefit/Cost
24	*Wastewater Quality	**Others**	
25	*Wastewater Pressure (Force Main)	58	*Customer Complaint
26	*Pipe Hydraulics	59	*Chemical
27	*Pipe Surcharging	60	*FOG
28	*Operational & Maintenance Practices	61	Resident Population Served
29	*Pipe Renewal Record	62	Failing Utilities
30	Pipe Failure Record	63	Consequence/Risk
31	*Infiltration/Inflow	64	Third-party Damage
32	*Exfiltration		
33	*Blockage/Stoppage		
34	*Sediments		
35	*Inspection Record		

Note: * Essential Parameter for Gold Standard

Condition Assessment and Rehabilitation Practice

The Town of Blacksburg prioritizes repair/rehabilitation/replacement of the wastewater collection system with these criteria:

- Age of the pipe
- Infiltration/inflow problems
- Surge problem
- Immediate attention requirements

During wet weather events, the Town of Blacksburg experienced overflows and surcharges in its wastewater collection system that caused violations of the. Clean Water Act. The Town recently contacted local consultants to conduct a sanitary sewer study. The objectives of this study were to assess the capacity and the overall condition of the existing wastewater collection system to determine if any portions of the system need repair, replacement, or rehabilitation. Additionally, the status of the wastewater collection system was assessed to determine if the system is in compliance with current and pending regulatory requirements [99]. The three aspects of this study are:

a) Capacity

Assessment of the wastewater collection system's capacity allows the town to determine where problems currently and potentially exist. The two capacity issues to be assessed are overflows and surcharging. Overflows can be spotted by visual inspection, while surcharging can be detected by manhole inspection and flow meter.

b) Condition

Cracked and broken pipes, leaking joints, sags, cracks and holes in the manhole walls, leakage in the manhole joints, and leakage through manhole joint frames are problems found in poor collection systems that cause inflow and infiltration, and reduce capacity. Also, some of these condition defects will reduce the capacity of the system from a structural standpoint. For example, solids accumulation in sags and partial flow blockage from broken pipes reduce the cross sectional area and decrease their capacities.

c) Current and Pending Regulations

The study involves research of the current and pending state and federal regulations applicable to wastewater collection systems. The historical data including previous studies was reviewed to identify the problem areas for rehabilitation decisions.

CHAPTER 8.0

FUTURE RESEARCH WORK

Chapter Highlights

This chapter presents the future research work for development of pipe prediction models:

- **Condition Assessment:** Many utilities have adopted WRc and NASSCO coding systems; however the existing coding systems are based only on CCTV inspection.

- **Standard Rating System:** The research team believes that there is a value in development of a robust and standard pipe condition rating system.

- **Data Analysis:** Thorough review and analysis of data and data mining are required to find the statistical significance of each parameter in the data model.

- **Condition Index:** Development of the condition/performance index for the wastewater pipe system will then lead to the development of prediction/deterioration models for wastewater pipe systems.

- **Deterioration Modeling:** It is a crucial step in assessing the remaining service life. Sound infrastructure deterioration models are essential for accurately predicting future conditions that, in turn, are key inputs to effective maintenance, repair and rehabilitation decision making.

- **Prediction Model:** Development of protocol and methodologies for predicting remaining economic life of pipes that will consist of three major components:
  1. Development of semi-empirical, mechanistic, and heuristic models.
  2. Development of experimental, probabilistic, and other mathematical models.
  3. Validation of proposed models through field data.

The research team identified and developed life cycles of wastewater pipe, failure modes and mechanisms, consequences of failure, data structures, and data collection protocols and methodologies. Future research work will focus on detailed analysis of knowledge and data acquired from participating utilities and pipe associations. The medium-term (Phase 2, August 2008-September 2009) research project will develop a standard procedure for rating the performance/condition of wastewater pipes. The long-term (Phase 3, August 2009-September

2010) research project will develop advanced artificial intelligence techniques and/or mathematical modeling methodologies for pipe deterioration models.

8.1 Development of Standard Performance Ratings Scale for Wastewater Pipes: Phase 2 Research Work (August 2008-September 2009)

Performance models will eventually play a key role in the wastewater infrastructure system as the performance or pavement models in pavement management systems directly influence the results of pavement management analysis [100]. These models also provide confirmation on the effectiveness of different design alternatives and help in the reduction of long-term maintenance and rehabilitation cost [101]. Overall the condition rating index (CRI) evaluates the pavement surface longitudinal deviation from a true planer surface, and is also known as roughness. The range for CRI is 1 to 100, where 100 mean an excellent pavement and 1 means the worst pavement. The expert's knowledge is used to provide relative weightings to the variables in the CRI evaluation model. Figure 8-1 shows an example of condition rating index, which ranges from 0 to 100 and each color represents the condition of the pipe.

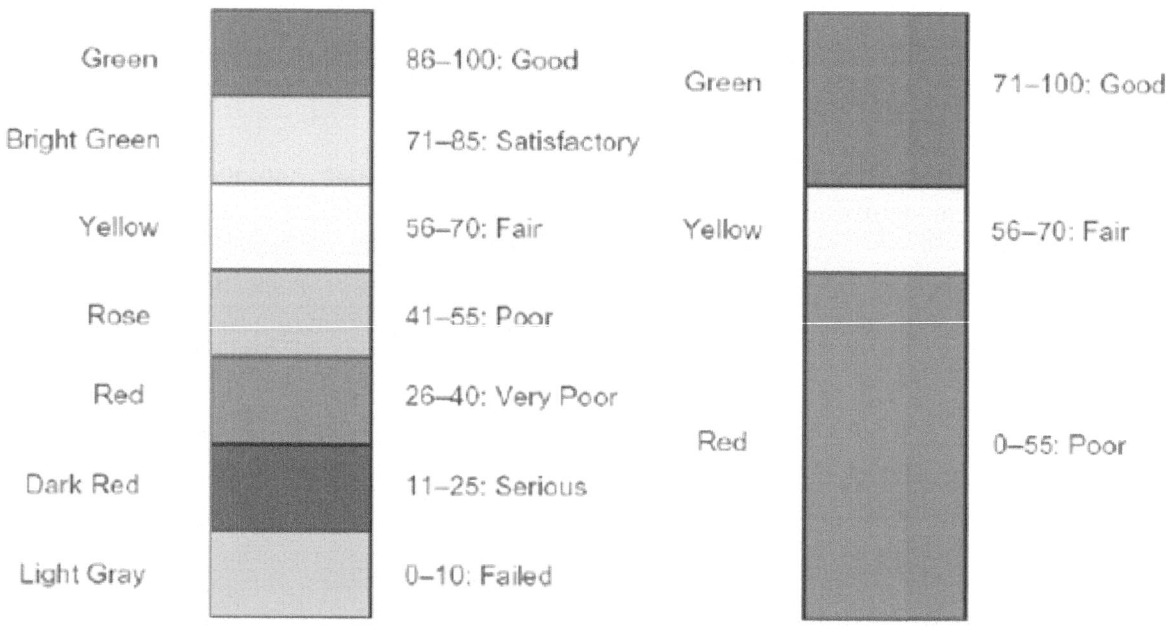

Figure 8-1. Condition Rating Index.

Many utilities have adopted WRc and NASSCO coding systems described in Chapter 2.0; however these coding systems are based on CCTV inspection of wastewater pipes.

The research team believes that a robust and standard condition rating system should be developed for wastewater pipes. To obtain a robust condition index for pipe infrastructure, all the identified parameters for the wastewater pipe system shown in Figure 8-2 should be mathematically combined to form condition indexes.

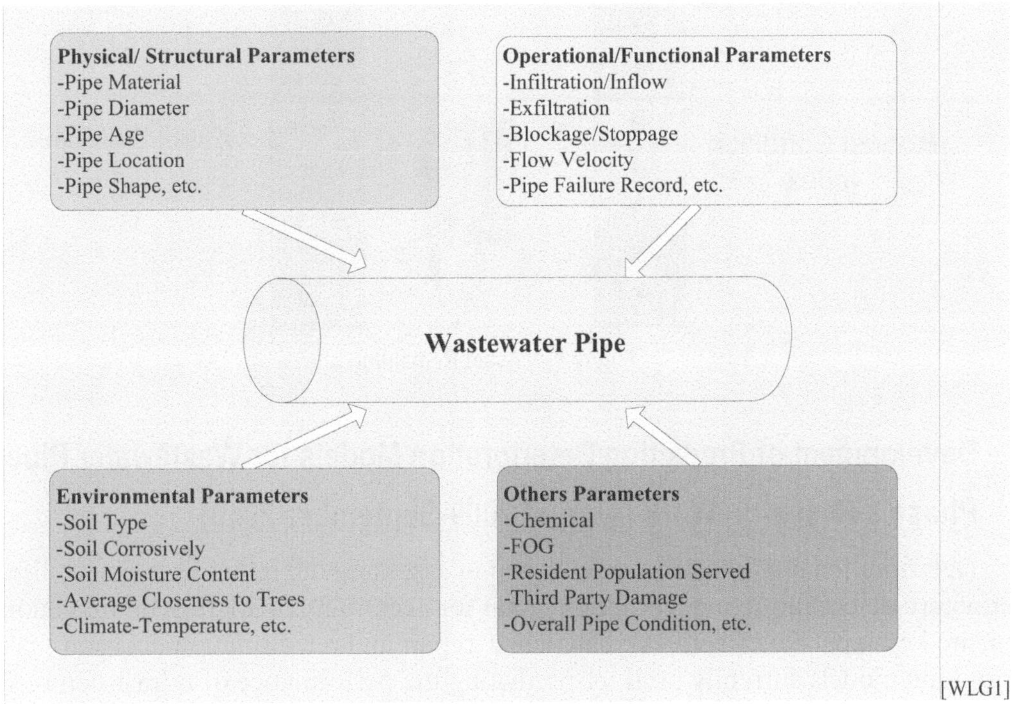

Figure 8-2. Wastewater Pipe Parameters.

Each component of the pipe is rated in terms of structural, functional, environmental, and other factors. An estimate of the overall condition can then be developed by combining the condition index of each functional system through a weighted summation that reflects the importance of the various factors. An example of proposed approach is presented in Figure 8-3.

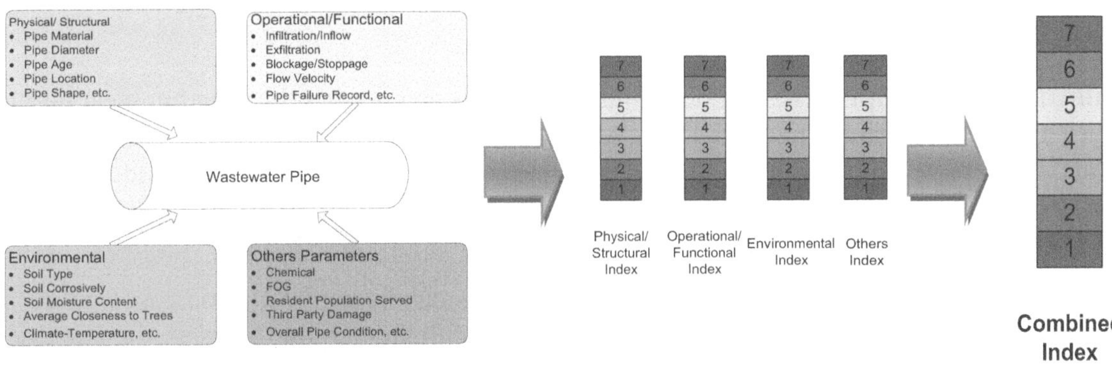

Figure 8-3. Development of Condition Rating Index.

The anticipated condition rating system will take every parameter from the developed data structure into account, considering not only the pipe but the entire system. The value of each parameter is not equivalent depending on circumstance and location. The research team must carefully examine this issue. Data analysis is required to find the statistical significant of each parameter. The robust condition scale can easily be linked to the asset remaining life for pipe infrastructure systems, as shown in Figure 8-4. The development of the condition/performance index for wastewater pipe systems will then lead to the prediction/deterioration models.

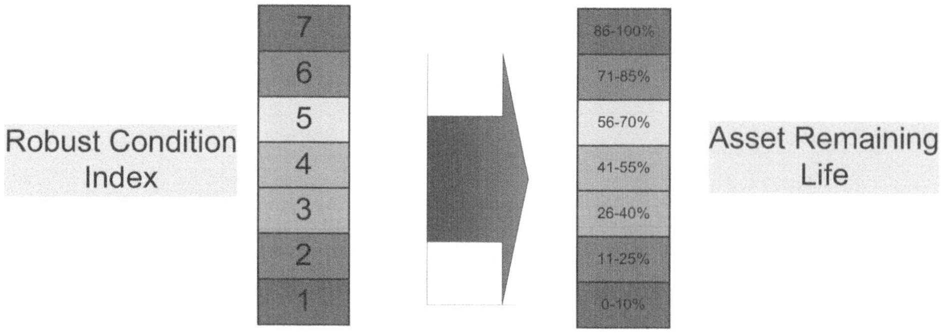

Figure 8-4. Asset Remaining Life.

8.2 Development of Prediction/Deterioration Models for Wastewater Pipes: Phase 3 Research Work (August 2009-September 2010)

Deterioration modeling is a crucial step in assessing the remaining service life. Sound infrastructure deterioration models are essential for accurately predicting future conditions that, in turn, are key inputs to effective maintenance, repair and rehabilitation decision making. Deterioration models currently used for predicting the performance of infrastructure systems include: straight line extrapolation, regression techniques and the probability-based Markovian model [99]. Deterioration models can be grouped into the following five major categories:

1. Deterministic models
2. Statistical models
3. Probabilistic models
4. Advanced mathematical models
5. Heuristic models

Deterministic Model Approach: The deterministic models can predict the remaining economic life of buried pipes. Deterministic models can be classified into the two classes: a) mechanistic models and b) empirical models. Some of the existing models in the transportation industry have been developed through regression analysis and combined mechanistic-empirical analysis. Data collection and lab/field experimental tests are required to develop these models.

Statistical Model Approach: Historical data-based models can be applied to forecast future failure rates of infrastructure. For example, an exponential function could represent the increase in water main failure rates with pipe age [103].

Probabilistic Model Approach: The probabilistic models can predict a distribution and range of values for dependent variables, such as the condition of buried pipes state vectors. Deterioration models for pipe infrastructure were developed using the Markov chains approach and data from sewer pipe inspection reports [104-107].

Advanced Mathematical Model Approach: Artificial neural networks (ANN) and fuzzy logic approaches have been used for infrastructure deterioration. Achim et al. [108] developed an ANN to predict the average failure rate in cast iron pipes. Kleiner et al. [109] used fuzzy logic to model deterioration and failure in pre-stressed concrete cylinder pipes and cast/ductile iron pipes.

Heuristic Model Approach: The heuristic models are more common for large and poorly understood infrastructure problems and produces sub-optimal solutions. These models can be developed through subjective opinion from experienced field engineers and experts.

8.2.1 Previous Work on Deterioration Models for Wastewater Pipes

Deterioration models for pipe infrastructure were developed using the Markov chains approach and data from sewer pipe inspection reports [68, 104, 105, 110]. For a reliable application of the Markov chain approach to deterioration forecasting, reliable assessment of the current condition is very important, because if the base information is incorrect, then the forecasting model based on that information will be unreliable [111].

Unfortunately, these models have not accounted for the presence of heterogeneity in data, which may lead to biased coefficient estimates. Furthermore, researchers have usually imposed a Markovian specification in the development of such models, implying that the probabilistic determination in a given period is independent of history [112].

Wirahadikusumah et al. [106] developed the deterioration model for American combined sewer pipes using a Markov model in conjunction with nonlinear optimization in 2001. Another Markov model developed in Australia by Micevski et al. [113] for the structural deterioration of stormwater pipes. The model is calibrated using Bayesian techniques. The parameters found to influence pipe deterioration process are diameter, construction material, soil type, and exposure classification. Baik et al. [114] summarized various studies attempt to develop the deterioration models for wastewater systems. Ariaratnam et al. [115] proposed a method for predicting the likelihood that a particular infrastructure system is in a deficient state, using logistic regression models, a special case of linear regression. A cohort survival model was developed by Rostum et al. [116], based on the condition assessment data, in this case CCTV. The wastewater systems were grouped based on construction period, pipe material, pipe dimensions, soil condition and etc., in this model. In 2006, Baik et al. [114] presented a probability model approach showing some advantages in estimating transition probabilities over the approaches developed in the past, including the nonlinear optimization-based method.

Recently, advanced mathematical methods have been used to predict pipeline failure. Achim et al. [117] used neural networks to predict water pipe deterioration. Tran et al. [108] developed artificial neural networks models to predict the stormwater pipe deterioration. The model was calibrated by Bayesian weight estimation using Markov Chain Monte Carlo simulation to account for the uncertainty.

8.2.2 Development of Pipe Deterioration Prediction Models

The deterioration modeling task for the pipe infrastructure will consist of three major components: 1) development of semi-empirical, mechanistic, and heuristic models; 2) development of experimental, probabilistic, and other mathematical models; and 3) validation of proposed models through observed data. An example of the development process is presented in Figure 8-5.

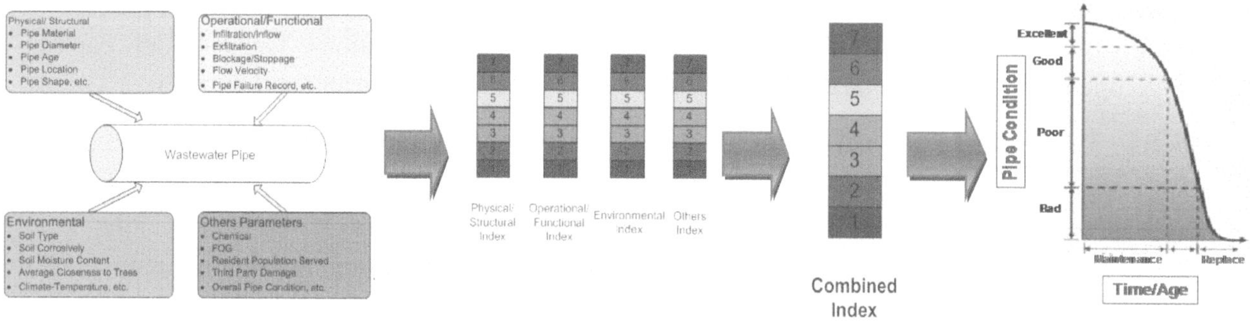

Figure 8-5. Development of Robust Pipe Deterioration Prediction Model.

An analysis approach will provide utility managers with a practical and efficient technique for the prediction of wastewater pipe performance. Researchers will also explore advanced artificial intelligence techniques and/or other mathematical modeling methodologies for deterioration models. The developed methodology will be evaluated using a time dependent data set of wastewater pipe network from various participating utilities across the nation.

As shown in Figure 8-5, the deterioration model is produced from the condition index moving over time. The deterioration curve should be updated when the actual condition of the pipe has been found. An example of described updated curve is presented in Figure 8-6.

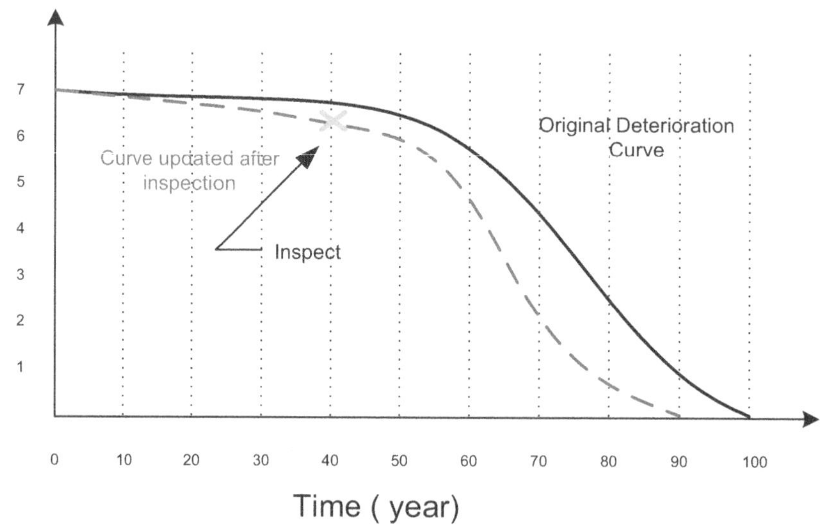

Figure 8-6. Updated Prediction/Deterioration Curve after Inspection.

Figure 8-7 illustrates the condition/performance prediction model. The left diagram shows pipe condition deteriorated over time and the right diagram shows the updated-condition of the pipe over time when the actions such as maintenance or rehabilitation has been performed.

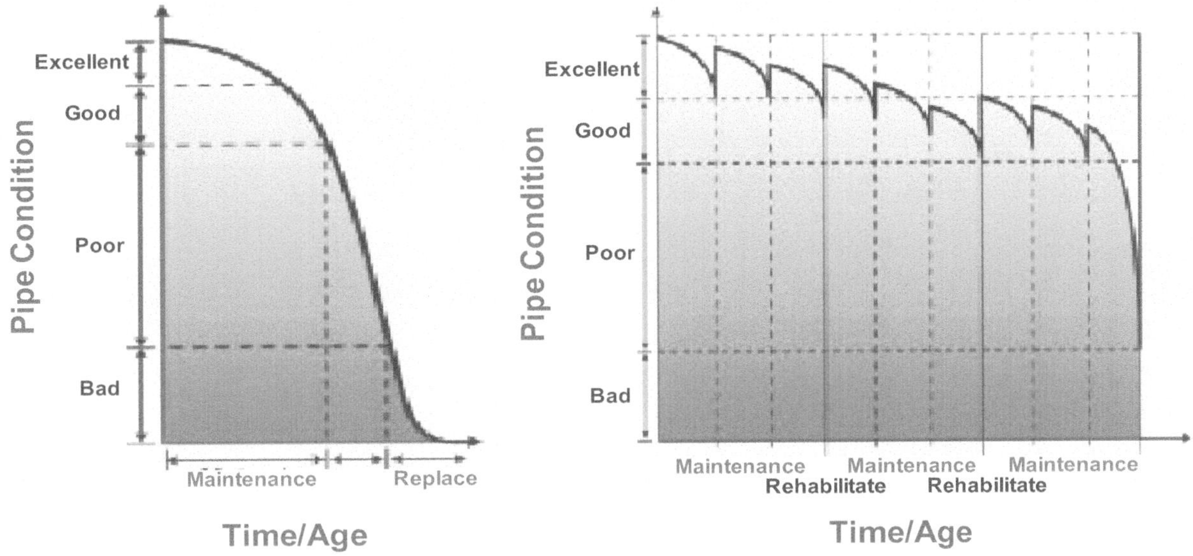

Figure 8-7. Condition/Performance Deterioration Prediction Model.

To develop a robust wastewater pipe deterioration prediction model, the research will attempt to capture and develop the deterioration curve based on environmental, soil, structural, hydraulic, and other factors. After understanding the deterioration process based on various factors, the research will develop an aggregate pipe infrastructure deterioration curve. An example of our proposed approach is shown in Figure 8-8.

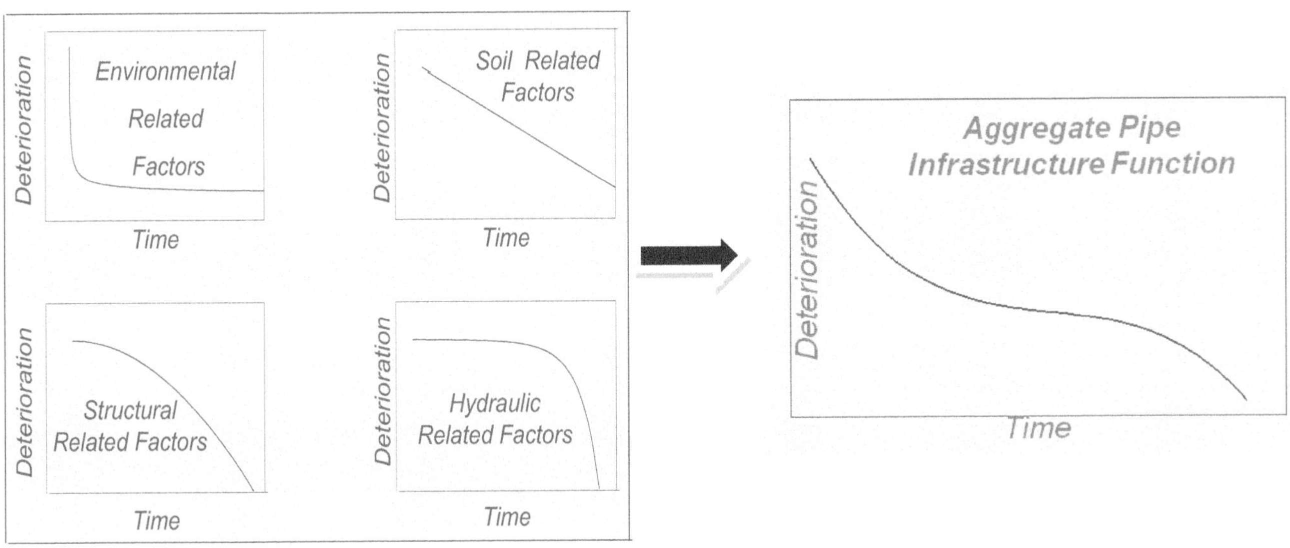

Figure 8-8. Development of Combined Wastewater Pipe Deterioration Curve.

Appendix A

WASTEWATER PIPE LIFE CYCLE

Appendix A-1

LIFE CYCLE OF CONCRETE PIPE

RCP and PCCP Wastewater Pipes Life Cycle			
Failure Causes			**Brief Explanation**
Design			
1	Poor Design		
		Connection	Connections not properly sized to fit pipe
		Vertical Connections	Connection types not properly selected
		Mechanical Vulnerable Material	Materials with erodible mortars
		Chemical Vulnerable Material	Materials vulnerable to sulphate attack
		Trust Restraint	Thrust restraint poorly designed
		Corrosion Control	Corrosion control method not properly selected
		Thickness	Thickness not properly designed for specified internal and external loads
		Special Applications	Applications such as pipe on supports poorly designed
2	Poor Project Planning		
		Improper Connections	Poor connections at joint
		Under-Design for Load	Excessive crushing or ring stress
		Time Limited Design	Designed for a short age expectancy
		Poor Assumptions of Environment	Poor support design, seasonal shrinkage/swelling, and migration of fines into or along bedding
3	Dimensioning		
Manufacture			
4	Manufacturing Defects		
		Wall Thickness	Thickness can vary around the circumference of pipe while manufacturing
		Inclusions of Unintended Structures	Inclusion in metal which are unintended
		Porosity due to Air Pockets	Air pockets trapped while manufacturing
		Longitudinal Surface Defects	Surface defects while manufacturing
		Cheap Composite Material	Materials with material that doesn't meet the standard
		Improper Connections	Poor connections at joint
5	Storage		
6	Observation and Quality Control		
Construction			
7	Transit		
		Transit Damage	Damage while in transit, or loading and unloading
8	Human Error		
		Third Party Damage	Excavations occur near pipeline
9	Poor Workmanship		
		Improper Connections	Poor connections at joint
		Uneven Bedding	Poor backfill operations
		Poor Backfilling Material	Corrosive backfill material
		Impact Damage	Excessive crushing
		Lining Issues	Lining installed incorrectly or poorly
10	Observation and Quality Control		
Operation & Maintenance			
11	Mechanical		
		Material Properties	Material changes over time and design strength is specified for certain time period
		Hydraulic Factors	Losses, thrust caused by internal pressure, water hammer, erosion of water pipe from soil, negative pressures
12	Thermal		
		Temperature Failures	Components subjected to temperatures above or below specification of pipe
13	Chemical		
		Chemical Attacks	Exposed to chemical substances
14	Biological		
		Biological Attacks	Anaerobic/Aerobic processes
15	External Interference		
		Movement of Soil	External pressure which is not symmetrical, possible loads
		Tensile and Compression Failures	Ductile tensile, brittle tensile, fatigue, bending, and compression failure
16	Internal Interference		
		Fatigue	Internal pressure, surge pressure
17	Natural Catastrophes		
18	Inappropriate Service		
19	Inappropriate Maintenance		
Repair/Rehabilitation/Replacement			
20	Design Error		
21	Manufacturing Error		
22	Construction Error		
23	Operation & Maintenance Error		

Appendix A-2

LIFE CYCLE OF CLAY PIPE

Clay Wastewater Pipe Life Cycle			
Failure Causes			**Brief Explanation**
Design			
1	Poor Design		
		Thickness	Thickness not properly designed for specified external loads
2	Poor Project Planning		
		Under-Design for Load	Excessive crushing or ring stress
		Poor Assumptions of Environment	Poor support design, seasonal shrinkage/swelling, and migration of fines into or along bedding
3	Dimensioning		
Manufacture			
4	Manufacturing Defects		
		Porosity due to Air Pockets	Air pockets trapped while manufacturing
5	Storage		
6	Observation and Quality Control		
Construction			
7	Transit		
		Transit Damage	Damage while in transit, or loading and unloading
8	Human Error		
		Third Party Damage	Excavations occur near pipeline
9	Poor Workmanship		
		Bell Hole	No bell hole, or bell hole is too small can cause bell cracking
		Uneven Bedding	Poor backfill operations
		Poor Backfilling Material	Backfill material doesn't meet the standard
		Impact Damage	Excessive crushing
10	Observation and Quality Control		
Operation & Maintenance			
11	External Interference		
		Movement of Soil	External pressure which is not symmetrical, possible loads
12	Differential Settlement		
13	Natural Catastrophes		
14	Inappropriate Service		
15	Inappropriate Maintenance		
Repair/Rehabilitation/Replacement			
16	Design Error		
17	Manufacturing Error		
18	Construction Error		
19	Operation & Maintenance Error		

Appendix A-3

LIFE CYCLE OF DUCTILE IRON (DI) PIPE

Ductile Iron Wastewater Pipe Life Cycle			
Failure Causes			**Brief Explanation**
Design			
1	Poor Design		
		Coating	Not designed for a thick enough coating
		Connection	Connections not properly sized to fit pipe
		Vertical Connections	Connection types not properly selected
		Chemical Vulnerable Material	Materials vulnerable to sulphate attack and corrosion
		Trust Restraint	Thrust restraint poorly designed
		Corrosion Control	Corrosion control method not properly selected
		Thickness	Thickness not properly designed for specified internal and external loads
		Special Applications	Applications such as pipe on supports poorly designed
2	Poor Project Planning		
		Improper Connections	Poor connections at joint
		Under-Design for Load	Excessive crushing or ring stress
		Time Limited Design	Designed for a short age expectancy
		Poor Assumptions of Environment	Poor support design, seasonal shrinkage/swelling, and migration of fines into or along bedding
3	Dimensioning		
Manufacture			
4	Manufacturing Defects		
		Wall Thickness	Thickness can vary around the circumference of pipe while manufacturing
		Inclusions of Unintended Structures	Inclusion in metal which are unintended
		Surface Defects	Surface defects during manufacturing
		Cheap Composite Material	Materials with material that doesn't meet the standard
		Improper Connections	Poor connections at joint
5	Storage		
6	Observation and Quality Control		
Construction			
7	Transit		
		Coating Damage	Damage to coating while in transit, or loading and unloading
8	Human Error		
		Third Party Damage	Excavations occur near pipeline
9	Poor Workmanship		
		Improper Connections	Poor connections at joint
		Uneven Bedding	Poor backfill operations
		Poor Backfilling Material	Corrosive backfill material
		Impact Damage	Excessive crushing
		Lining Issues	Lining installed incorrectly or poorly
10	Observation and Quality Control		
Operation & Maintenance			
11	Mechanical		
		Material Properties	Material changes over time and design strength is specified for certain time period
		Hydraulic Factors	Losses, thrust caused by internal pressure, water hammer, erosion of water pipe from soil, negative pressures
12	Thermal		
		Temperature Failures	Components subjected to temperatures above or below specification of pipe
13	Chemical		
		Pitting Corrosion	Corrosion in one particular area
		Galvanic Corrosion	Reaction of dissimilar metals
		Graphitization	Mass of residual graphite flakes interspersed with oxides of iron
14	Biological		
		Microbiologically Influenced Corrosion	Anaerobic/Aerobic corrosion
15	External Interference		
		Dissimilar Electrolytes	Changes in ground current or if pipe is exposed to two different soil types
		Stray Current Corrosion	Stray direct currents, interference currents, stray alternating currents
		Movement of Soil	External pressure which is not symmetrical, possible loads
		Tensile and Compression Failures	Ductile tensile, brittle tensile, fatigue, bending, and compression failure
16	Internal Interference		
		Fatigue	caused by internal pressure, surge pressure
17	Natural Catastrophes		

Failure Causes		Brief Explanation
18	Inappropriate Service	
19	Inappropriate Maintenance	
Repair/Rehabilitation/Replacement		
20	Design Error	
21	Manufacturing Error	
22	Construction Error	
23	Operation & Maintenance Error	

Appendix A-4

LIFE CYCLE OF CAST IRON (CI) PIPE

Grey Cast Iron Wastewater Pipe Life Cycle			
Failure Causes			**Brief Explanation**
Design			
1	Poor Design		
		Coating	Not designed for a thick enough coating
		Connection	Connections not properly sized to fit pipe
		Vertical Connections	Connection types not properly selected
		Mechanical Vulnerable Material	materials vulnerable to erosion
		Chemical Vulnerable Material	Materials vulnerable to sulphate attack and corrosion
		Trust Restraint	Thrust restraint poorly designed
		Corrosion Control	
		Thickness	Thickness not properly designed for specified internal and external loads
		Special Applications	Applications such as pipe on supports poorly designed
2	Poor Project Planning		
		Improper Connections	Poor connections at joint
		Under-Design for Load	Excessive crushing or ring stress
		Time Limited Design	Designed for a short age expectancy
		Poor Assumptions of Environment	Poor support design, seasonal shrinkage/swelling, and migration of fines into or along bedding
3	Dimensioning		
Manufacture			
4	Manufacturing Defects		
		Wall Thickness	Thickness can vary around the circumference of pipe while cast
		Inclusions of Unintended Structures	Inclusion in metal which are unintended
		Porosity due to Air Pockets	Air pockets trapped while casting
		Longitudinal Surface Defects	Surface defects due to mold while casting
		Cheap Composite Material	Materials with material that doesn't meet the standard
		Pipe Material Content	Networks cause when phosphorus is added to molten cast iron
		Improper Connections	Poor connections at joint
5	Storage		
6	Observation and Quality Control		
Construction			
7	Transit		
		Coating Damage	Damage to coating while in transit, or loading and unloading
8	Human Error		
		Third Party Damage	Excavations occur near pipeline
9	Poor Workmanship		
		Improper Connections	Poor connections at joint
		Uneven Bedding	Poor backfill operations
		Poor Backfilling Material	Corrosive backfill material
		Impact Damage	Excessive crushing
		Lining Issues	Lining installed incorrectly or poorly
10	Observation and Quality Control		
Operation & Maintenance			
11	Mechanical		
		Material Properties	Material changes over time and design strength is specified for certain time period
		Hydraulic Factors	Losses, thrust caused by internal pressure, water hammer, erosion of water pipe from soil, negative pressures
12	Thermal		
		Temperature Failures	Components subjected to temperatures above or below specification of pipe
13	Chemical		
		Pitting Corrosion	Corrosion in one particular area
		Galvanic Corrosion	Reaction of dissimilar metals
		Graphitization	Mass of residual graphite flakes interspersed with oxides of iron
14	Biological		
		Microbiologically Influenced Corrosion	Anaerobic/Aerobic corrosion
15	External Interference		
		Dissimilar Electrolytes	Changes in ground current or if pipe is exposed to two different soil types
		Stray Current Corrosion	Stray direct currents, interference currents, stray alternating currents
		Movement of Soil	External pressure which is not symmetrical, possible loads
		Tensile and Compression Failures	Ductile tensile, brittle tensile, fatigue, bending ,and compression failure

Failure Causes			Brief Explanation
16	Internal Interference		
		Fatigue	Caused by internal pressure, surge pressure
17	Natural Catastrophes		
18	Inappropriate Service		
19	Inappropriate Maintenance		
Repair/Rehabilitation/Replacement			
20	Design Error		
21	Manufacturing Error		
22	Construction Error		
23	Operation & Maintenance Error		

Appendix A-5

LIFE CYCLE OF POLYVINYL CHLORIDE (PVC) PIPE

PVC Wastewater Pipe Life Cycle			
Failure Causes			**Brief Explanation**
Design			
1	Poor Design		
		Connection	Connections not properly sized to fit pipe
		Vertical Connections	Connection types not properly selected
		Trust Restraint	Thrust restraint poorly designed
		Thickness	Thickness not properly designed for specified internal and external loads
		Special Applications	Applications such as pipe on supports poorly designed
2	Poor Project Planning		
		Improper Connections	Poor connections at joint
		Under-Design for Load	Excessive crushing or ring stress
		Time Limited Design	Designed for a short age expectancy
		Poor Assumptions of Environment	Poor support design, seasonal shrinkage/swelling, and migration of fines into or along bedding
3	Dimensioning		
Manufacture			
4	Manufacturing Defects		
		Wall Thickness	Thickness can vary around the circumference of pipe while manufacturing
		Inclusions of Unintended Structures	Inclusion in material which are unintended
		Porosity due to Air Pockets	Air pockets trapped while manufacturing
		Surface Defects	Surface defects during manufacturing
		Cheap Composite Material	Materials with material that doesn't meet the standard
		Improper Connections	Poor connections at joint
5	Storage		
6	Observation and Quality Control		
Construction			
7	Transit		
		Transit Damage	Damage while in transit, or loading and unloading
8	Human Error		
		Third Party Damage	Excavations occur near pipeline
9	Poor Workmanship		
		Improper Connections	Poor connections at joint
		Uneven Bedding	Backfill material doesn't meet the standard
		Poor Backfilling Material	Corrosive backfill material
		Impact Damage	Excessive crushing
10	Observation and Quality Control		
Operation & Maintenance			
11	Mechanical		
		Material Properties	Material changes over time and design strength is specified for certain time period
		Hydraulic Factors	Losses, thrust caused by internal pressure, water hammer, erosion of water pipe from soil, negative pressures
12	Thermal		
		Temperature Failures	Components subjected to temperatures above or below specification of pipe
13	Chemical		
		Chemical Attacks	Exposed to chemical substances
14	Biological		
		Biological Attacks	Anaerobic/Aerobic processes
15	External Interference		
		Movement of Soil	External pressure which is not symmetrical, possible loads
		Tensile and Compression Failures	Ductile tensile, brittle tensile, fatigue, bending, and compression failure
16	Internal Interference		
		Fatigue	Internal pressure, surge pressure
17	Natural Catastrophes		
18	Inappropriate Service		
19	Inappropriate Maintenance		
Repair/Rehabilitation/Replacement			
20	Design Error		
21	Manufacturing Error		
22	Construction Error		
23	Operation & Maintenance Error		

Appendix A-6

Life Cycle of Polyethylene (PE) and High Density Polyethylene (HDPE) Pipes

PE and HDPE Wastewater Pipes Life Cycle			
Failure Causes			**Brief Explanation**
Design			
1	Poor Design		
		Connection	Connections not properly sized to fit pipe
		Vertical Connections	Connection types not properly selected
		Trust Restraint	Thrust restraint poorly designed
		Thickness	Thickness not properly designed for specified external loads
		Special Applications	Applications such as pipe on supports poorly designed
2	Poor Project Planning		
		Improper Connections	Poor connections at joint
		Under-Design for Load	Excessive crushing or ring stress
		Time Limited Design	Designed for a short age expectancy
		Poor Assumptions of Environment	Poor support design, seasonal shrinkage/swelling, and migration of fines into or along bedding
3	Dimensioning		
Manufacture			
4	Manufacturing Defects		
		Wall Thickness	Thickness can vary around the circumference of pipe while manufacturing
		Inclusions of Unintended Structures	Inclusion in material which are unintended
		Porosity due to Air Pockets	Air pockets trapped while manufacturing
		Surface Defects	Surface defects during manufacturing
		Cheap Composite Material	Materials with material that doesn't meet the standard
		Improper Connections	Poor connections at joint
5	Storage		
6	Observation and Quality Control		
Construction			
7	Transit		
		Transit Damage	Damage while in transit, or loading and unloading
8	Human Error		
		Third Party Damage	Excavations occur near pipeline
9	Poor Workmanship		
		Improper Connections	Poor connections at joint
		Uneven Bedding	Backfill material doesn't meet the standard
		Poor Backfilling Material	Corrosive backfill material
		Impact Damage	Excessive crushing
10	Observation and Quality Control		
Operation & Maintenance			
11	Mechanical		
		Material Properties	Material changes over time and design strength is specified for certain time period
12	Thermal		
		Temperature Failures	Components subjected to temperatures above or below specification of pipe
13	Chemical		
		Chemical Attacks	Exposed to chemical substances
14	Biological		
		Biological Attacks	Anaerobic/Aerobic processes
15	External Interference		
		Movement of Soil	External pressure which is not symmetrical, possible loads
		Tensile and Compression Failures	Ductile tensile, brittle tensile, fatigue, bending , compression failure
16	Internal Interference		
		Fatigue	Internal pressure, surge pressure
17	Natural Catastrophes		
18	Inappropriate Service		
19	Inappropriate Maintenance		
Repair/Rehabilitation/Replacement			
20	Design Error		
21	Manufacturing Error		
22	Construction Error		
23	Operation & Maintenance Error		

Appendix A-7

LIFE CYCLE OF BRICK SEWER

Brick Sewers Life Cycle			
Failure Causes			**Brief Explanation**
Design			
1	Poor Design		
		Mechanical Vulnerable Material	Materials with erodible mortars
		Chemical Vulnerable Material	Materials vulnerable to sulphate attack
		Wall Thickness	Thickness not properly designed for and external loads
2	Poor Project Planning		
		Under-Design for Load	Excessive crushing or ring stress
		Time Limited Design	Designed for a short age expectancy
		Poor Assumptions of Environment	Poor support design, seasonal shrinkage/swelling, and migration of fines into or along bedding
3	Dimensioning		
Construction			
4	Human Error		
		Third Party Damage	Excavations occur near pipeline
5	Poor Workmanship		
		Uneven Bedding	Poor backfill operations
		Impact Damage	Excessive crushing
6	Observation and Quality Control		
Operation & Maintenance			
7	Thermal		
		Temperature Failures	Components subjected to temperatures above or below average
8	Chemical		
		Chemical Attacks	Exposed to chemical substances
9	Biological		
		Biological Attacks	Anaerobic/Aerobic processes
10	External Interference		
		Movement of Soil	External pressure which is not symmetrical, possible loads
		Tensile and Compression Failures	Ductile tensile, brittle tensile, fatigue, bending, and compression failure
11	Natural Catastrophes		
12	Inappropriate Service		
13	Inappropriate Maintenance		
Repair/Rehabilitation/Replacement			
14	Design Error		
15	Manufacturing Error		
16	Construction Error		
17	Operation & Maintenance Error		

Appendix B

WASTEWATER PIPE FAILURE MODES AND MECHANISMS

Appendix B-1

GRAVITY CONCRETE PIPE (CP)

Non-Reinforced Concrete (CP) Wastewater Pipes Failure Modes and Mechanisms (Gravity)		
A. Main Line		
Failure Mode		Mechanism
Typical Failure Mode		
Structural Failure		
1	**Crack**	
	Longitudinal crack	Excessive crushing or ring stress from non-uniform bedding support and ground movement, may occur at springing level, crown, and invert
	Circumferential crack	Vertical movement a result of excessive shear and bending stress, most likely occur near joint, differential settlement
	Multiple cracks	Combination of stresses that cause longitudinal and circumferential cracks, impact during construction
2	**Fracture**	
	Longitudinal fracture	Excessive crushing or ring stress from non-uniform bedding support differential settlement and ground movement, may occur at springing level, crown, and invert
	Circumferential fracture	Vertical movement a result of excessive shear and bending stress, most likely occur near joint
	Multiple fractures	Combination of stresses that cause longitudinal and circumferential fractures
3	**Broken**	Occurs when pieces of a cracked or fractured pipe move form their original position
4	**Hole**	Broken pipe has completely dislodged from the wall, or corrosion
5	**Deformed**	
	Vertical deformed	Occurs when a fractured or broken pipe loses the support of the surrounding ground, overloading
	Horizontal deformed	Occurs when a fractured or broken pipe loses the support of the surrounding ground
6	**Collapsed**	A further stage of deterioration after deformation
7	**Surface damage**	
	Roughness increased	Hydrogen Sulfide attack, chemical erosion, abrasion
	Aggregate visible	Hydrogen Sulfide attack, chemical erosion, abrasion
	Aggregate projecting	Hydrogen Sulfide attack, chemical erosion, abrasion
	Aggregate missing	Hydrogen Sulfide attack, chemical erosion, abrasion
	Missing wall	Hydrogen Sulfide attack, chemical erosion, abrasion
	Surface spalling	Movement of pipe, a chipped pipe, or the expansion action of corroded reinforcement
8	**Lining Failure**	
	Detached	Poor installation, manufacturing defects, maintenance damage, impact damage during installation
	Defective End	Poor installation, manufacturing defects, maintenance damage, impact damage during installation
	Blistered	Poor installation, manufacturing defects, maintenance damage, impact damage during installation
	Service Cur Shifted	Poor installation, manufacturing defects, maintenance damage, impact damage during installation
	Abandoned Connection	Poor installation
	Overcut Service	Poor installation, liners shrink after the services are cut
	Undercut Service	Poor installation
	Buckled	Poor installation, manufacturing defects, maintenance damage, impact damage during installation
	Wrinkled	Poor installation, manufacturing defects, maintenance damage, impact damage during installation
	Spall	Poor installation, manufacturing defects, maintenance damage, impact damage during installation
	Crack	Poor installation, manufacturing defects, maintenance damage, impact damage during installation
Operational and Maintenance Failure		
9	**Deposits**	
	Attached	foreign materials get into the pipe. Normally be concentrated along misaligned joints or fractures
	Settled	flat grade, deposits are often distributed throughout a pipe
	Ingress	Surrounding materials conveyed into pipe through the infiltration of groundwater
10	**Roots**	
	Fine	Roots ingress through pipe to defects; such as holes and cracks in the pipe
	Tap	Over half inch thick roots enter barrel of pipe through defects
	Medium	Many roots have accumulated and formed a mess
	Ball	Next step of root intrusion from medium, roots have formed a mess
11	**Infiltration**	
	Weeper	Slow ingress of water through defective joint or pipe wall
	Dripper	Water dripping through defective joint or pipe wall
	Runner	Water running into sewer pipe through defective joint or pipe wall
	Gusher	Under pressure, water entering pipe through defects or joint
12	**Exfiltration**	Holes, cracks and/or external loading
13	**Obstacles/Obstructions**	
	Pipe material in invert	Large or medium sections of broken pipe lying in the invert
	Object intruding through wall	Third party objects inserting through the pipe wall after the pipeline construction
	Object wedged in the joint	Object is wedged in the joint of pipeline
	Object through connection	Other objects entering into the main pipe via the connection and obstructed at the pipe connection
	External pipe or cable	Third party objects inserting through the pipe wall after the pipeline construction
	Construction debris	Construction debris that is left in the pipeline
	Rocks	Rocks obstructing the flow
	Other Obstacles	Other objects have obstructed the flow
Rare Failure Mode		
Structural Failure		
14	**Crack**	
	Spiral cracks	Combination of stresses that cause longitudinal and circumferential cracks
15	**Fracture**	
	Spiral fractures	Combination of stresses that cause longitudinal and circumferential fractures

Concrete Pipe (CP) Wastewater Pipes Failure Modes and Mechanisms (Gravity)		
B. Joint		
Failure Mode		Mechanism
Typical Failure Mode		
1	**Lined Joint**	
	Separated Joint	Ground movement, poor bedding, differential settlement
2	**Rubber Gasket Joint**	
	Corrosion degradation of reinforcing wires	Corrosion, hydrogen sulfide attack
	Gasket degradation	Aging, chemical attack
	Concrete degradation	soil corrosivity
	Mortar seal degradation	Chemical attack
	Seal displacement	Ground movement and poor bedding
3	**Steel Endring Joint**	
	Weld failure	Poor welding, poor filling material
	Corrosion of spigot	leak Joint, failed seal, hydrogen sulfide attack
	Ovality of pipe	Excessive loads, bending stress, differential settlement

Appendix B-2

GRAVITY REINFORCE CONCRETE (RCP) PIPE

Non Reinforced Concrete (CP) Wastewater Pipes Failure Modes and Mechanisms (Gravity)

A. Main Line

Failure Mode		Mechanism
Typical Failure Mode		
Structural Failure		
1	**Crack**	
	Longitudinal crack	Excessive crushing or ring stress from non-uniform bedding support and ground movement, may occur at springing level, crown, and invert
	Circumferential crack	Vertical movement a result of excessive shear and bending stress, most likely occur near joint, differential settlement
	Multiple cracks	Combination of stresses that cause longitudinal and circumferential cracks, impact during construction
2	**Fracture**	
	Longitudinal fracture	Excessive crushing or ring stress from non-uniform bedding support differential settlement and ground movement, may occur at springing level, crown, and invert
	Circumferential fracture	Vertical movement a result of excessive shear and bending stress, most likely occur near joint
	Multiple fractures	Combination of stresses that cause longitudinal and circumferential fractures
3	**Broken**	Occurs when pieces of a cracked or fractured pipe move form their original position
4	**Hole**	Broken pipe has completely dislodged from the wall, or corrosion
5	**Deformed**	
	Vertical deformed	Occurs when a fractured or broken pipe loses the support of the surrounding ground, overloading
	Horizontal deformed	Occurs when a fractured or broken pipe loses the support of the surrounding ground
6	**Collapsed**	A further stage of deterioration after deformation
7	**Surface damage**	
	Roughness increased	Hydrogen Sulfide attack, chemical erosion, abrasion
	Aggregate visible	Hydrogen Sulfide attack, chemical erosion, abrasion
	Aggregate projecting	Hydrogen Sulfide attack, chemical erosion, abrasion
	Aggregate missing	Hydrogen Sulfide attack, chemical erosion, abrasion
	Reinforcement visible	Hydrogen Sulfide attack, chemical erosion, abrasion
	Reinforcement projecting	Hydrogen Sulfide attack, chemical erosion, abrasion
	Reinforcement corroded	Hydrogen Sulfide attack, chemical erosion, abrasion, Exposure to corrosive substances, soil corrosivity, stray current
	Missing wall	Hydrogen Sulfide attack, chemical erosion, abrasion
	Surface spalling	Movement of pipe, a chipped pipe, or the expansion action of corroded reinforcement
8	**Lining Failure**	
	Detached	Poor installation, manufacturing defects, maintenance damage, impact damage during installation
	Defective End	Poor installation, manufacturing defects, maintenance damage, impact damage during installation
	Blistered	Poor installation, manufacturing defects, maintenance damage, impact damage during installation
	Service Cur Shifted	Poor installation, manufacturing defects, maintenance damage, impact damage during installation
	Abandoned Connection	Poor installation
	Overcut Service	Poor installation, liners shrink after the services are cut
	Undercut Service	Poor installation
	Buckled	Poor installation, manufacturing defects, maintenance damage, impact damage during installation
	Wrinkled	Poor installation, manufacturing defects, maintenance damage, impact damage during installation
	Spall	Poor installation, manufacturing defects, maintenance damage, impact damage during installation
	Crack	Poor installation, manufacturing defects, maintenance damage, impact damage during installation
Operational and Maintenance Failure		
9	**Deposits**	
	Attached	foreign materials get into the pipe. Normally be concentrated along misaligned joints or fractures
	Settled	flat grade, deposits are often distributed throughout a pipe
	Ingress	Surrounding materials conveyed into pipe through the infiltration of groundwater
10	**Roots**	
	Fine	Roots ingress through pipe to defects; such as holes and cracks in the pipe
	Tap	Over half inch thick roots enter barrel of pipe through defects
	Medium	Many roots have accumulated and formed a mess
	Ball	Next step of root intrusion from medium, roots have formed a mess
11	**Infiltration**	
	Weeper	Slow ingress of water through defective joint or pipe wall
	Dripper	Water dripping through defective joint or pipe wall
	Runner	Water running into sewer pipe through defective joint or pipe wall
	Gusher	Under pressure, water entering pipe through defects or joint
12	**Exfiltration**	Holes, cracks and/or external loading
13	**Obstacles/Obstructions**	
	Pipe material in invert	Large or medium sections of broken pipe lying in the invert
	Object intruding through wall	Third party objects inserting through the pipe wall after the pipeline construction
	Object wedged in the joint	Object is wedged in the joint of pipeline
	Object through connection	Other objects entering into the main pipe via the connection and obstructed at the pipe connection
	External pipe or cable	Third party objects inserting through the pipe wall after the pipeline construction
	Construction debris	Construction debris that is left in the pipeline
	Rocks	Rocks obstructing the flow
	Other Obstacles	Other objects have obstructed the flow
Rare Failure Mode		
Structural Failure		
14	**Crack**	
	Spiral cracks	Combination of stresses that cause longitudinal and circumferential cracks
15	**Fracture**	
	Spiral fractures	Combination of stresses that cause longitudinal and circumferential fractures

RCP Wastewater Pipes Failure Modes and Mechanisms (Gravity)		
B. Joint		
Failure Mode		**Mechanism**
Typical Failure Mode		
1	**Lined Joint**	
	Separated Joint	Ground movement, poor bedding, differential settlement
2	**Rubber Gasket Joint**	
	Corrosion degradation of reinforcing wires	Corrosion, hydrogen sulfide attack
	Gasket degradation	Aging, chemical attack
	Concrete degradation	soil corrosivity
	Mortar seal degradation	Chemical attack
	Seal displacement	Ground movement and poor bedding
3	**Steel Endring Joint**	
	Weld failure	Poor welding, poor filling material
	Corrosion of spigot	leak Joint, failed seal, hydrogen sulfide attack
	Ovality of pipe	Excessive loads, bending stress, differential settlement

Appendix B-3

GRAVITY CLAY PIPE

Clay Wastewater Pipe Failure Modes and Mechanisms *A. Main Line*		
Failure Mode		Mechanism
Typical Failure Mode		
Structural Failure		
1	**Cracks / Fractures**	Excessive crushing and/or ring stress results from non-uniform bedding support, differential settlement and ground movement, damage during transportation
	Longitudinal (Localized) crushing failure	Poor construction (no bell hole), poor design, not adequate longitudinal support, poor bedding
	Star breaks	excessive point loading
	Shear Failure	Uneven support between adjacent pipe
	Lateral Shear	Construction problems
	Shear between manhole pipeline	differential settlement
2	**Collapsed**	external loading, ground movement, differential settlement, poor bedding
Operational and Maintenance Failure		
3	**Sags in the line**	Poor installation (angular deflection greater that 5%), ground movement, instable foundation
4	**Deposits**	
	Attached	foreign materials get into the pipe. Normally be concentrated along misaligned joints
	Settled	flat grade, deposits are often distributed throughout a pipe
	Ingress	Surrounding materials conveyed into pipe through the infiltration of groundwater
5	**Roots**	
	Fine	Roots ingress through pipe structural damages
	Tap	Over half inch thick roots enter barrel of pipe through damages
	Medium	Many roots have accumulated and formed a mess
	Ball	Next step of root intrusion from medium, roots have formed a mess
6	**Infiltration**	
	Weeper	Slow ingress of water through damaged joint or pipe wall
	Dripper	Water dripping through damaged joint or pipe wall
	Runner	Water running into sewer pipe through damaged joint or pipe wall
	Gusher	Under pressure, water entering pipe through damages or joint
7	**Exfiltration**	water ingresses through damages or separated joints to environment
Rare Failure Mode		
Structural Failure		
8	**Surface Damage**	Chemical attacks (Concentrated Hydrofluoric acid, 200 degrees Sodium Hydroxide)
9	**Abrasion**	accumulate of abrasive materials, such as rocks, and sand
Operational and Maintenance Failure		
10	**Obstacles/Obstructions**	
	Pipe material in invert	Large or medium sections of broken pipe lying in the invert
	Object intruding through wall	Third party objects inserting through the pipe wall after the pipeline construction
	Object wedged in the joint	Object is wedged in the joint of pipeline
	Object through connection	Other objects entering into the main pipe via the connection and obstructed at the pipe connection
	External pipe or cable	Third party objects inserting through the pipe wall after the pipeline construction
	Construction debris	Construction debris that is left in the pipeline
	Rocks	Rocks obstructing the flow
	Other Obstacles	Other objects have obstructed the flow

Clay Wastewater Pipe Failure Modes and Mechanisms *B. Joint*		
Failure Mode		Mechanism
Typical Failure Mode		
1	**Parallel offsets**	Did not have a self-sealing joint, tolerances between spigot and bell ends were loose, ground movement
2	**Pulled Joints**	Poor installation, ground movement and not proper bedding condition
3	**Excessive angular deflection**	improper installation, ground movement, poor bedding
4	**Joint Leakage**	Installation (Dirt on the joint during installation)

Appendix B-4

GRAVITY DUCTILE IRON (DI) PIPE

Ductile Iron Wastewater Pipe Failure Modes and Mechanisms (Gravity) *A. Main Line*		
Failure Mode		Mechanism
Typical Failure Mode		
Structural Failure		
1	**Hole**	Corrosion
2	**Corrosion**	
	External Corrosion	Exposure to corrosive substances, soil corrosivity, stray current
	Internal Corrosion	Exposure to corrosive substances, wastewater characteristics
3	**Deformed**	
	Vertical Deformed	pipe loses the support of the surrounding ground, overloading, impact loading
	Horizontal Deformed	pipe loses the support of the surrounding ground, overloading, impact loading
4	**Lining Failure**	
	Detached	Poor installation, manufacturing defects, maintenance damage, impact damage during installation
	Defective End	Poor installation, manufacturing defects, maintenance damage, impact damage during installation
	Blistered	Poor installation, manufacturing defects, maintenance damage, impact damage during installation
	Service Cur Shifted	Poor installation, manufacturing defects, maintenance damage, impact damage during installation
	Abandoned Connection	Poor installation
	Overcut Service	Poor installation, liners shrink after the services are cut
	Undercut Service	Poor installation
	Buckled	Poor installation, manufacturing defects, maintenance damage, impact damage during installation
	Wrinkled	Poor installation, manufacturing defects, maintenance damage, impact damage during installation
	Spall	Poor installation, manufacturing defects, maintenance damage, impact damage during installation
	Crack	Poor installation, manufacturing defects, maintenance damage, impact damage during installation
Operational and Maintenance Failure		
5	**Deposits**	
	Attached	foreign materials get into the pipe. Normally be concentrated along misaligned joints or fractures
	Settled	flat grade, deposits are often distributed throughout a pipe
	Ingress	Surrounding materials conveyed into pipe through the infiltration of groundwater
6	**Roots**	
	Fine	Roots ingress through pipe defects; such as holes and cracks in the pipe
	Tap	Over half inch thick roots enter barrel of pipe through defects
	Medium	Many roots have accumulated and formed a mess
	Ball	Next step of root intrusion from medium, roots have formed a mess
7	**Infiltration**	
	Weeper	Slow ingress of water through defective joint or pipe wall
	Dripper	Water dripping through defective joint or pipe wall
	Runner	Water running into sewer pipe through defective joint or pipe wall
	Gusher	Under pressure water entering pipe through defective joint or pipe wall
8	**Exfiltration**	Holes, cracks
Rare Failure Mode		
Structural Failure		
9	**Crack**	
	Circumferential Crack	Thermal contraction, poor support leading to movement, external loading
	Longitudinal Crack	crushing forces acting along the pipe
	Spiral Crack	Combination of bending stress and hoop stress from internal pressure
	Mixed Crack	Combination of bending stress and hoop stress from internal pressure
10	**Fracture**	
	Circumferential Fracture	Thermal contraction, poor support leading to movement, external loading
	Longitudinal Fracture	Internal water pressure, crushing forces acting along the pipe
	Spiral Fracture	Combination of bending stress and hoop stress from internal pressure
	Mixed Fracture	Combination of bending stress and hoop stress from internal pressure
	Shearing	
	Bell Shearing	Bending forces and/or compressive forces pushing the spigot of the pipe into the bell of the next pipeline
11	**Collapsed**	excessive loading, third party damage
Operational and Maintenance Failure		
12	**Obstacles/Obstructions**	
	Object Intruding Through Wall	Third party objects inserting through the pipe wall after the pipeline construction
	Object Wedged in the Joint	Object is wedged in the joint of pipeline
	Object Through Connection	Other objects entering into the main pipe via the connection and obstructed at the pipe connection
	External Pipe or Cable	Third party objects inserting through the pipe wall after the pipeline construction
	Other Obstacles	Other objects have obstructed the flow

Ductile Iron Wastewater Pipe Failure Modes and Mechanisms (Gravity) *B. Joint*		
Failure Mode		Failure Mechanism
Typical Failure Modes		
1	**Push-On**	
	Gasket/Seal Failure	Installation, stray currents, corrosive ground
	Angular Deflection	Installation, ground movement, poor bedding
	Axial Pull Out	Installation, ground movement
	Thermal Expansion/Cracked Spigot	Installation, high temperature, freezing-thawing
	Lateral Pipe Movement	Installation, ground movement
	Internal Corrosion	Installation, exposure to corrosive substances
	External Corrosion	Installation, corrosive ground, stray currents
2	**Bolted Flange**	
	Bolt Failure	Installation, stray currents, corrosive ground
	Flange Failure	Installation, stray currents, corrosive ground
	Gasket Failure	Installation, stray currents, corrosive ground
	Stress at Flange	Installation, deflection of the pipe
	Bolt Corrosion	Installation, stray currents, corrosive ground
3	**Restrained Joint**	
	Pull Out	Installation, ground movement
	External Corrosion	Installation, stray currents, corrosive ground
Rare Failure Modes		
4	**Mechanical Joint**	
	Bolt Failure	Installation, stray currents, corrosive ground
	Bolt Corrosion	Installation, stray currents, corrosive ground
	Backing Ring Corrosion	Installation, stray currents, corrosive ground
	Structural Spigot Failure	Installation, stray currents, corrosive ground
5	**Slip-On Coupler**	
	Internal Corrosion	Installation, expose to corrosive substances
	External Corrosion	Installation, stray currents, corrosive ground
	Bolt Corrosion	Installation, stray currents, corrosive ground
	Spigot Pull-Out	Installation, ground movement
6	**Ball & Socket**	
	Structural Socket Failure	Installation, stray currents

Ductile Iron Wastewater Pipe Failure Modes and Mechanisms (Gravity) *C. Lateral*		
Failure Mode		**Mechanism**
Typical Failure Mode		
Structural Failure		
1	**Hole**	Corrosion
2	**Corrosion**	Exposure to corrosive substances, soil corrosivity, stray current
3	**Deformed**	pipe loses the support of the surrounding ground, overloading, impact loading
4	**Lining Failure**	Poor installation, manufacturing defects, maintenance damage, impact damage during installation
Operational and Maintenance Failure		
5	**Deposits**	flat grade, foreign material get into the pipe
6	**Roots**	Roots ingress through pipe defects; such as holes and cracks in the pipe
7	**Infiltration**	water entering pipe through defective connection or pipe wall
8	**Exfiltration**	Holes, cracks
Rare Failure Mode		
Structural Failure		
9	**Crack**	External forces acting along the pipe, ground movement
10	**Fracture**	External forces acting along the pipe, ground movement
11	**Collapsed**	excessive loading, third party damage
Operational and Maintenance Failure		
12	**Obstacles/Obstructions**	Third party objects entering into the pipe during or after construction

Appendix B-5

GRAVITY CAST IRON (CI) PIPE

Cast Iron Wastewater Pipe Failure Modes and Mechanisms (Gravity) *A. Main Line*		
Failure Mode		Mechanism
Typical Failure Mode		
Structural Failure		
1	**Crack**	
	Circumferential Crack	Thermal contraction, poor support leading to movement, external loading
	Longitudinal Crack	Crushing forces acting along the pipe, impact loading
	Spiral Crack	Combination of bending stress and hoop stress from internal pressure
	Mixed Crack	Combination of bending stress and hoop stress from internal pressure
2	**Fracture**	
	Circumferential Fracture	Thermal contraction, poor support leading to movement, external loading
	Longitudinal Fracture	Crushing forces acting along the pipe, impact loading
	Spiral Fracture	Combination of bending stress and hoop stress from internal pressure
	Mixed Fracture	Combination of bending stress and hoop stress from internal pressure
3	**Hole**	
	Blowout Holes	Corrosion pitting, casting flaws
4	**Corrosion**	
	External Corrosion	Exposure to corrosive substances, soil corrosivity, stray current
	Internal Corrosion	Exposure to corrosive substances, wastewater characteristics
5	**Lining Failure**	
	Spall	Poor installation, manufacturing defects, maintenance damage, impact damage during installation
	Crack	Poor installation, manufacturing defects, maintenance damage, impact damage during installation
Operational and Maintenance Failure		
6	**Deposits**	
	Attached	Foreign materials get into the pipe. Normally be concentrated along misaligned joints or fractures
	Settled	Flat grade, deposits are often distributed throughout a pipe
	Ingress	Surrounding materials conveyed into pipe through the infiltration of groundwater
7	**Roots**	
	Fine	Roots ingress through pipe to defects; such as holes and cracks in the pipe
	Tap	Over half inch thick roots enter barrel of pipe through defects
	Medium	Many roots have accumulated and formed a mess
	Ball	Next step of root intrusion from medium, roots have formed a mess
8	**Infiltration**	
	Weeper	Slow ingress of water through defective joint or pipe wall
	Dripper	Water dripping through defective joint or pipe wall
	Runner	Water running into sewer pipe through defective joint or pipe wall
	Gusher	Under pressure, water entering pipe through defects or joint
9	**Exfiltration**	Holes, cracks caused by internal pressure and/or external loading
Rare Failure Mode		
Structural Failure		
10	**Shearing**	
	Bell Shearing	Bending forces and/or compressive forces pushing the spigot of the pipe into the bell of the next
11	**Burst Failure**	
	Burst Failure (Int.)	Internal pitting and graphitization corrosion weakening wall often combined with induced strains
	Burst Failure (Ext.)	External pitting and graphitization corrosion weakening wall often combined with induced strains
	Burst Failure (3rd party)	Third party impact damage
12	**Collapsed**	Excessive loading, extremely high pressure, vacuum loading, third party damage
Operational and Maintenance Failure		
13	**Obstacles/Obstructions**	
	Object Intruding Through Wall	Third party objects inserting through the pipe wall after the pipeline construction
	Object Wedged in the Joint	Object is wedged in the joint of pipeline
	Object Through Connection	Other objects entering into the main pipe via the connection and obstructed at the pipe connection
	External Pipe or Cable	Third party objects inserting through the pipe wall after the pipeline construction
	Other Obstacles	Other objects have obstructed the flow

Cast Iron Wastewater Pipe Failure Modes and Mechanisms (Gravity) *B. Joint*		
Failure Mode		**Failure Mechanism**
Typical Failure Modes		
1	**Mechanical Joint**	
	Bolt Failure	Installation, stray currents, corrosive ground
	Bolt Corrosion	Installation, stray currents, corrosive ground
	Backing Ring Corrosion	Installation, stray currents, corrosive ground
	Structural Spigot Failure	Installation, stray currents, corrosive ground
2	**Bolted Flange**	
	Bolt Failure	Installation, stray currents, corrosive ground
	Flange Failure	Installation, stray currents, corrosive ground
	Gasket Failure	Installation, stray currents, corrosive ground
	Stress at Flange	Installation, deflection of the pipe
	Bolt Corrosion	Installation, stray currents, corrosive ground
3	**Slip-On Coupler**	
	Internal Corrosion	Installation, expose to corrosive substances
	External Corrosion	Installation, stray currents, corrosive ground
	Bolt Corrosion	Installation, stray currents, corrosive ground
	Spigot Pull-Out	Installation, ground movement, water hammer
4	**Groove & Shoulder**	
	Structural Socket Failure	Installation, stray currents
5	**Ball & Socket**	
	Structural Socket Failure	Installation, stray currents
Rare Failure Modes		
6	**Push-On**	
	Gasket/Seal Failure	Installation, stray currents, corrosive ground
	Angular Deflection	Installation, ground movement, poor bedding
	Axial Pull Out	Installation, internal water pressure, water hammer
	Thermal Expansion/Cracked Spigot	Installation, high temperature, freezing-thawing
	Lateral Pipe Movement	Installation, ground movement, water hammer
	Internal Corrosion	Installation, exposure to corrosive substances
	External Corrosion	Installation, corrosive ground, stray currents
7	**Ball & Spigot Lead/Jute**	
	Angular Deflection	Installation, stray currents
	Axial Pull Out	Installation, stray currents
	Lead/Jute Deterioration	Installation, stray currents
	Thermal Expansion/Cracked Spigot	Installation, stray currents
	Lateral Pipe Movement	Installation, stray currents
	External Corrosion	Installation, stray currents

Cast Iron Wastewater Pipe Failure Modes and Mechanisms (Gravity) *C. Lateral*		
Failure Mode		Mechanism
Typical Failure Mode		
Structural Failure		
1	Crack	Thermal contraction, poor support leading to movement, external loading
2	Fracture	Thermal contraction, poor support leading to movement, external loading
3	Hole	Corrosion pitting, casting flaws
4	Corrosion	Exposure to corrosive substances, soil corrosivity, stray current
5	Lining Failure	Poor installation, manufacturing defects, maintenance damage, impact damage during installation
Operational and Maintenance Failure		
6	Deposits	Foreign materials get into the pipe, flat grade
7	Roots	Roots ingress through pipe to defects; such as holes and cracks in the pipe
8	Infiltration	Water entering pipe through damaged joint or pipe wall
9	Exfiltration	Holes, cracks caused by internal pressure and/or external loading
Rare Failure Mode		
Structural Failure		
10	Shearing	Bending forces and/or compressive forces pushing the spigot of the pipe into the bell of the next
11	Burst Failure	Pitting and graphitization corrosion weakening wall often combined with induced strains
12	Collapsed	excessive loading, extremely high pressure, vacuum loading, third party damage
Operational and Maintenance Failure		
13	Obstacles/Obstructions	Third party objects inserting through the pipe wall during and/or after the pipeline construction

Appendix B-6

Gravity Polyvinyl Chloride (PVC) Pipe

PVC Wastewater Pipe Failure Modes and Mechanisms (Gravity)

A. Main Line

Failure Mode		Mechanism
Typical Failure Mode		
Structural Failure		
1	**Crack**	
	Longitudinal crack	Crushing forces acting along the pipe, bending
	Spiral crack	Combination of bending stress and hoop stress
	Bell Splitting	Over insertion, excessive axial alignment,
2	**Fracture**	
	Longitudinal fracture	Excessive crushing or ring stress from non-uniform bedding support differential settlement and ground movement
	Spiral fracture	Combination of ring and axial stresses
3	**Broken**	Occurs when pieces of a cracked or fractured pipe move form their original position
4	**Hole**	Broken pipe has completely dislodged from the wall
5	**Deformed**	
	Axial	External pressure, settlement, uplift, over insertion and interventions
6	**Buckling**	
	Axial Buckling	External pressure, axial compression, temperatures, fire, and interventions
	Transverse/Ring Buckling	External pressure, axial compression, production, residual stresses, high temperatures, fire, and interventions
	Non-Symmetric Buckling	Longitudinal bending and brazier effect
	Longitudinal Buckling	Axial compression and thermal effects
	Other	External pressure, axial compression, production, residual stresses, high temperatures, fire, and interventions
7	**Color Change**	UV radiation, chemicals, biological, and excessive stresses
8	**Blisters/Voids**	
	Voids	Production and chemicals
Operational and Maintenance Failure		
9	**Deposits**	
	Attached	foreign materials get into the pipe. Normally be concentrated along misaligned joints or fractures
	Settled	flat grade, deposits are often distributed throughout a pipe
	Ingress	Surrounding materials conveyed into pipe through the infiltration of groundwater
10	**Roots**	
	Fine	Roots ingress through pipe defects; such as holes and cracks in the pipe
	Tap	Over half inch thick roots enter barrel of pipe through defects
	Medium	Many roots have accumulated and formed a mess
	Ball	Next step of root intrusion from medium, roots have formed a mess
11	**Infiltration**	
	Weeper	Slow ingress of water through defective joint or pipe wall
	Dripper	Water dripping through defective joint or pipe wall
	Runner	Water running into sewer pipe through defective joint or pipe wall
	Gusher	Under pressure water entering pipe through defective joint or pipe wall
12	**Exfiltration**	Holes, cracks and/or external loading
Rare Failure Mode		
Structural Failure		
13	**Crack**	
	Circumferential crack	Thermal contraction, poor support leading to movement
	Mixed Cracking	Production, impact, vibration, fatigue, residual stresses, UV radiation, frost, chemicals, biological, and interventions
	Irregular Crack	Environmental and stress cracking, pipe surface embrittled by UV, Chemical attacks
	Rapid Crack Propagation (RCP)	Dynamic loading at low temperatures
14	**Fracture**	
	Circumferential fracture	Vertical movement a result of excessive shear and bending stress
	Mixed fractures	Combination of stresses that cause longitudinal and circumferential cracking
15	**Crazing**	
	Through Crazing	Axial tension, settlement, fatigue, chemicals, and residual stresses, bad manufacturing
	External Crazing	Axial tension, settlement, fatigue, chemicals, and residual stresses, bad manufacturing
	Internal Crazing	Axial tension, settlement, fatigue, chemicals, and residual stresses, bad manufacturing
16	**Deformed**	
	Ring	External pressure, axial tension, axial compression, bending, traffic load, settlement, uplift, high temperatures, fire, and interventions
17	**Dimension Change**	High temperatures, chemicals, and fire
18	**Blisters/Voids**	
	External Blister	Excessive amount of heat
	Internal Blister	Production, chemicals
	Blazes	Osmosis
	Delamination	Production, improper fusion (acid tone test)
19	**Abrasion/Erosion**	Flow velocity and interventions
20	**Chemical Attacks**	Expose to Chemical such as solvent or oil
21	**Collapse**	External loading, third party damage
22	**Other**	
	Blank Ends	attachment, not properly joint

Failure Mode		Mechanism
	Blown T	attachment, not properly joint
	Fire Plug Riser Split	attachment, not properly joint
	Ball Valve	attachment, not properly joint
	End Blown Off	attachment, not properly joint
Operational and Maintenance Failure		
23	**Obstacles/Obstructions**	
	Object intruding through wall	Third party objects inserting through the pipe wall after the pipeline construction
	Object wedged in the joint	Object is wedged in the joint of pipeline
	Object through connection	Other objects entering into the main pipe via the connection and obstructed at the pipe connection
	External pipe or cable	Third party objects inserting through the pipe wall after the pipeline construction
	Other Obstacles	Other objects have obstructed the flow

PVC Wastewater Pipe Failure Modes and Mechanisms (Gravity)		
B. Joint		
Failure Mode		Mechanism
Typical Failure Mode		
1	**Push-fit and slip-on collar (Gasket)**	
	Displaced gasket (into socket)	Poor installation, improper alignment
	Cracking of socket	Over insertion, axial deflecting joint
Rare Failure Modes		
2	**Push-fit and slip-on collar (Gasket)**	
	Spigot deflection	Deflection of the pipe, differential settlement
	Gasket degradation	Exposure to chemical (solvent, oil), UV
	Bell splitting	Surge fatigue, water hammer
	PVC material degradation	Exposure to chemical (solvent, oil), UV
3	**Solvent Weld**	
	Poor Adhesion degradation	Poor Adhesive
	Embrittlement	Temperature changes
	Bell splitting	Surge fatigue, water hammer
	Angular deflection	Ground movement, poor bedding
	Misalignment	Poor installation, ground movement
4	**Other**	
	Joint Leak	
	Mechanical Joint Leak	Improper Tightening
	Blown Rubber on Flange	
	Blown Joint	
	Leak on Collar	Gasket degradation
	Leaking Rubber Ring	Material degradation
	Gasket/Seal Failure	Displaced gasket into socket, gasket degradation
	Structural Bell Failure	Cracking of socket
	Butt Fusion Joint	Weak fusion weld under axial, shear, and bending forces
	Electro-Fusion Joint	Axial tension, bending, and improper welding
	Flange Joint	Improper welding, axial tension, and bending

PVC Wastewater Pipe Failure Modes and Mechanisms (Gravity) *C. Lateral*		
Failure Mode		Mechanism
Typical Failure Mode		
Structural Failure		
1	Crack	External forces acting along the pipe, ground movement
2	Fracture	External forces acting along the pipe, ground movement
3	Broken	Occurs when pieces of a cracked or fractured pipe move form their original position
4	Hole	Broken pipe has completely dislodged from the wall, chemical attacks
5	Deformed	External pressure, settlement, uplift, over insertion and interventions
6	Buckling	External pressure, axial compression, temperatures, fire, and interventions
7	Color Change	UV radiation, chemicals, biological, and excessive stresses
8	Blisters/Voids	Production and chemicals
Operational and Maintenance Failure		
9	Deposits	flat grade, foreign material get into the pipe
10	Roots	Roots ingress through pipe defects; such as holes and cracks in the pipe
11	Infiltration	water entering pipe through defective connection or pipe wall
12	Exfiltration	Holes, cracks
Rare Failure Mode		
Structural Failure		
14	Dimension Change	High temperatures, chemicals, and fire
16	Abrasion/Erosion	Flow velocity and interventions
17	Chemical Attacks	Expose to Chemical such as solvent or oil
18	Collapse	External loading, third party damage
Operational and Maintenance Failure		
19	Obstacles/Obstructions	Third party objects entering into the pipe during or after construction

Appendix B-7

Gravity Polyethylene (PE) and High Density Polyethylene (HDPE) Pipes

PE and HDPE Wastewater Pipe Failure Modes and Mechanisms (Gravity) *A. Main Line*		
Failure Mode		Mechanism
Typical Failure Mode		
Structural Failure		
1	**Crack**	
	Longitudinal crack	Crushing forces acting along the pipe, bending
	Spiral crack	Combination of ring and axial stresses
2	**Fracture**	
	Longitudinal fracture	Excessive crushing or ring stress from non-uniform bedding support differential settlement and ground movement
	Spiral fracture	Combination of ring and axial stresses
3	**Broken**	Occurs when pieces of a cracked or fractured pipe move form their original position
4	**Hole**	Broken pipe has completely dislodged from the wall
5	**Deformed**	
	Axial	External pressure, settlement, uplift, over insertion and interventions
Operational and Maintenance Failure		
6	**Deposits**	
	Attached	foreign materials get into the pipe. Normally be concentrated along misaligned joints or fractures
	Settled	flat grade, deposits are often distributed throughout a pipe
	Ingress	Surrounding materials conveyed into pipe through the infiltration of groundwater
7	**Roots**	
	Fine	Roots ingress through pipe defects; such as holes and cracks in the pipe
	Tap	Over half inch thick roots enter barrel of pipe through defects
	Medium	Many roots have accumulated and formed a mess
	Ball	Next step of root intrusion from medium, roots have formed a mess
8	**Infiltration**	
	Weeper	Slow ingress of water through defective joint or pipe wall
	Dripper	Water dripping through defective joint or pipe wall
	Runner	Water running into sewer pipe through defective joint or pipe wall
	Gusher	Under pressure water entering pipe through defective joint or pipe wall
9	**Exfiltration**	Holes, cracks and/or external loading
Rare Failure Mode		
Structural Failure		
10	**Crack**	
	Circumferential crack	Thermal contraction, poor support leading to movement
	Mixed Crack	Combination of stresses that cause longitudinal and circumferential cracking
	Irregular Crack	Environmental and stress cracking, pipe surface embrittled by UV, Chemical attacks
11	**Fracture**	
	Circumferential fracture	Vertical movement a result of excessive shear and bending stress
	Mixed fractures	Combination of stresses that cause longitudinal and circumferential cracking
12	**Deformed**	
	Ring	External pressure, axial tension, axial compression, bending, traffic load, settlement, uplift, high temperatures, fire, and interventions
13	**Dimension Change**	High temperatures, chemicals, and fire
14	**Abrasion/Erosion**	Flow velocity and interventions
15	**Chemical Attacks**	Expose to Chemical such as solvent or oil
16	**Collapse**	Excessive External loading, third party damage
Operational and Maintenance Failure		
17	**Obstacles/Obstructions**	
	Object intruding through wall	Third party objects inserting through the pipe wall after the pipeline construction
	Object wedged in the joint	Object is wedged in the joint of pipeline
	Object through connection	Other objects entering into the main pipe via the connection and obstructed at the pipe connection
	External pipe or cable	Third party objects inserting through the pipe wall after the pipeline construction
	Other Obstacles	Other objects have obstructed the flow

PE and HDPE Wastewater Pipe Failure Modes and Mechanisms (Gravity)		
B. Joint		
Failure Mode		Mechanism
Typical Failure Mode		
1	**Butt Fusion**	
	Poor joint integrity	Poor material used
	Gaps/cracks/splits/voids/dirt at weld int	Poor welding
	Mixed SDR materials	Poor installing practices
	Lateral offset or misalignment	Poor installation, ground movement
2	**Electrofusion**	
	Pull out of pipe in fitting	Ground movement, Pipe deflection
	Poor joint integrity	Poor material used
	Voids/contamination	Poor installation practices
	Overheating of fusion wire	Poor installation practices
3	**Compression fitting**	
	High local stress	high-pressure water
	Pull-out	Ground movement, Pipe deflection
	Cracking	Poor installation, ground movement
	Insert size - leakage	Poor installation practices

Appendix B-8

GRAVITY BRICK SEWER

Brick Sewer Failure Modes and Mechanisms		
Failure Mode		Mechanism
Typical Failure Mode		
Structural Failure		
1	**Crack**	External loading, poor bedding and supports
2	**Fracture**	External loading, poor bedding and supports
3	**Broken**	Occurs when pieces of a cracked or fractured pipe move form their original position
4	**Hole**	Broken brick has completely dislodged from the wall
5	**Deformed**	-
	Vertical deformed	Occurs when a fractured or broken sewer loses the support of the surrounding ground, overloading
	Horizontal deformed	Occurs when a fractured or broken sewer loses the support of the surrounding ground
6	**Collapsed**	A further stage of deterioration after deformation
7	**Surface spalling**	Movement of brick, lose of mortar
Operational and Maintenance Failure		
8	**Deposits**	
	Attached	foreign materials get into the pipe. Normally be concentrated along misaligned joints or fractures
	Settled	flat grade, deposits are often distributed throughout a pipe
	Ingress	Surrounding materials conveyed into pipe through the infiltration of groundwater
9	**Roots**	
	Fine	Roots ingress through pipe to defects; such as holes and cracks in the pipe
	Tap	Over half inch thick roots enter barrel of pipe through defects
	Medium	Many roots have accumulated and formed a mess
	Ball	Next step of root intrusion from medium, roots have formed a mess
10	**Infiltration**	
	Weeper	Slow ingress of water through defective joint or pipe wall
	Dripper	Water dripping through defective joint or pipe wall
	Runner	Water running into sewer pipe through defective joint or pipe wall
	Gusher	Under pressure, water entering pipe through defects or joint
11	**Exfiltration**	Holes, cracks and/or external loading
12	**Obstacles/Obstructions**	
	Pipe material in invert	Large or medium sections of broken pipe lying in the invert
	Object intruding through wall	Third party objects inserting through the pipe wall after the pipeline construction
	Object wedged in the joint	Object is wedged in the joint of pipeline
	Object through connection	Other objects entering into the main pipe via the connection and obstructed at the pipe connection
	External pipe or cable	Third party objects inserting through the pipe wall after the pipeline construction
	Construction debris	Construction debris that is left in the pipeline
	Rocks	Rocks obstructing the flow
	Other Obstacles	Other objects have obstructed the flow

Appendix B-9

Force Main Pre-Stressed Concrete Cylinder (PCCP) Pipe

PCCP Wastewater Pipe Failure Modes and Mechanisms (Force Main)

A. Main Line

Failure Mode		Mechanism
Typical Failure Mode		
Structural Failure		
1	**Crack**	
	Longitudinal crack	Excessive crushing or ring stress from non-uniform bedding support and ground movement, may occur at springing level, crown, and invert
	Circumferential crack	Vertical movement a result of excessive shear and bending stress, most likely occur near joint, differential settlement
	Multiple cracks	Combination of stresses that cause longitudinal and circumferential cracks, impact during construction
2	**Fracture**	
	Longitudinal fracture	Excessive crushing or ring stress from non-uniform bedding support differential settlement and ground movement, may occur at springing level, crown, and invert
	Circumferential fracture	Vertical movement a result of excessive shear and bending stress, most likely occur near joint
	Multiple fractures	Combination of stresses that cause longitudinal and circumferential fractures
3	**Broken**	Occurs when pieces of a cracked or fractured pipe move form their original position
4	**Hole**	Broken pipe has completely dislodged from the wall, or corrosion
5	**Deformed**	
	Vertical deformed	Occurs when a fractured or broken pipe loses the support of the surrounding ground, overloading
	Horizontal deformed	Occurs when a fractured or broken pipe loses the support of the surrounding ground
6	**Collapsed**	A further stage of deterioration after deformation
7	**Surface damage**	
	Roughness increased	Hydrogen Sulfide attack, chemical erosion, abrasion
	Aggregate visible	Hydrogen Sulfide attack, chemical erosion, abrasion
	Aggregate projecting	Hydrogen Sulfide attack, chemical erosion, abrasion
	Aggregate missing	Hydrogen Sulfide attack, chemical erosion, abrasion
	Reinforcement visible	Hydrogen Sulfide attack, chemical erosion, abrasion
	Reinforcement projecting	Hydrogen Sulfide attack, chemical erosion, abrasion
	Reinforcement corroded	Hydrogen Sulfide attack, chemical erosion, abrasion, Exposure to corrosive substances, soil corrosivity, stray current
	Missing wall	Hydrogen Sulfide attack, chemical erosion, abrasion
	Surface spalling	Movement of pipe, a chipped pipe, or the expansion action of corroded reinforcement
8	**Wire Break**	Combination of corroded reinforcement and high pressure
9	**Lining Failure**	
	Defective Weld	
	Detached	Poor installation, manufacturing defects, maintenance damage, impact damage during installation
	Defective end	Poor installation, manufacturing defects, maintenance damage, impact damage during installation
	Blistered	Poor installation, manufacturing defects, maintenance damage, impact damage during installation
	Buckled	Poor installation, manufacturing defects, maintenance damage, impact damage during installation
	Wrinkled	Poor installation, manufacturing defects, maintenance damage, impact damage during installation
Operational and Maintenance Failure		
10	**Exfiltration**	Holes, cracks and/or external loading
11	**Obstacles/Obstructions**	
	Object intruding through wall	Third party objects inserting through the pipe wall after the pipeline construction
	Object wedged in the joint	Object is wedged in the joint of pipeline
	Object through connection	Other objects entering into the main pipe via the connection and obstructed at the pipe connection
	External pipe or cable	Third party objects inserting through the pipe wall after the pipeline construction
	Other Obstacles	Other objects have obstructed the flow
Rare Failure Mode		
Structural Failure		
12	**Crack**	
	Spiral cracks	Combination of stresses that cause longitudinal and circumferential cracks
13	**Fracture**	
	Spiral fractures	Combination of stresses that cause longitudinal and circumferential fractures

PCCP Wastewater Pipes Failure Modes and Mechanisms (Force Main)		
B. Joint		
Failure Mode		**Mechanism**
Typical Failure Mode		
1	**Lined Joint**	
	Separated Joint	Ground movement, poor bedding, differential settlement
2	**Rubber Gasket Joint**	
	Corrosion degradation of reinforcing wires	Corrosion, hydrogen sulfide attack
	Gasket degradation	Aging, chemical attack
	Concrete degradation	soil corrosivity
	Mortar seal degradation	Chemical attack
	Seal displacement	Ground movement and poor bedding
3	**Steel Endring Joint**	
	Weld failure	Poor welding, poor filling material
	Corrosion of spigot	leak Joint, failed seal, hydrogen sulfide attack
	Ovality of pipe	Excessive loads, bending stress, differential settlement

Appendix B-10

FORCE MAIN DUCTILE IRON (DI) PIPE

Ductile Iron Wastewater Pipe Failure Modes and Mechanisms (Force Main)

A. Main Line

Failure Mode		Mechanism
Typical Failure Mode		
Structural Failure		
1	**Hole**	
	Blowout Holes	Corrosion pitting
2	**Corrosion**	
	External Corrosion	Exposure to corrosive substances, soil corrosivity, stray current
	Internal Corrosion	Exposure to corrosive substances, wastewater characteristics
3	**Deformed**	
	Vertical Deformed	pipe loses the support of the surrounding ground, overloading, impact loading
	Horizontal Deformed	pipe loses the support of the surrounding ground, overloading, impact loading
4	**Lining Failure**	
	Detached	Poor installation, manufacturing defects, maintenance damage, impact damage during installation
	Defective End	Poor installation, manufacturing defects, maintenance damage, impact damage during installation
	Blistered	Poor installation, manufacturing defects, maintenance damage, impact damage during installation
	Buckled	Poor installation, manufacturing defects, maintenance damage, impact damage during installation
	Wrinkled	Poor installation, manufacturing defects, maintenance damage, impact damage during installation
	Spall	Poor installation, manufacturing defects, maintenance damage, impact damage during installation
	Crack	Poor installation, manufacturing defects, maintenance damage, impact damage during installation
Operational and Maintenance Failure		
5	**Exfiltration**	Holes, cracks caused by internal pressure and/or external loading
Rare Failure Mode		
Structural Failure		
6	**Crack**	
	Circumferential Cracking	Thermal contraction, poor support leading to movement, external loading
	Longitudinal Crack	Internal water pressure, crushing forces acting along the pipe
	Spiral Crack	Combination of bending stress and hoop stress from internal pressure
	Mixed Cracking	Combination of bending stress and hoop stress from internal pressure
7	**Fracture**	
	Circumferential Fracture	Thermal contraction, poor support leading to movement, external loading
	Longitudinal Fracture	Internal water pressure, crushing forces acting along the pipe
	Spiral Fracture	Combination of bending stress and hoop stress from internal pressure
	Mixed Fracture	Combination of bending stress and hoop stress from internal pressure
	Shearing	
	Bell Shearing	Bending forces and/or compressive forces pushing the spigot of the pipe into the bell of the next pipeline
8	**Burst Failure**	
	Burst Failure (Int.)	Internal pitting and graphitization corrosion weakening wall often combined with induced strains
	Burst Failure (Ext.)	External pitting and graphitization corrosion weakening wall often combined with induced strains
	Burst Failure (3rd party)	Third party impact damage
9	**Collapsed**	excessive loading, extremely high pressure, vacuum loading, third party damage
Operational and Maintenance Failure		
10	**Obstacles/Obstructions**	
	Object Intruding Through Wall	Third party objects inserting through the pipe wall after the pipeline construction
	Object Wedged in the Joint	Object is wedged in the joint of pipeline
	Object Through Connection	Other objects entering into the main pipe via the connection and obstructed at the pipe connection
	External Pipe or Cable	Third party objects inserting through the pipe wall after the pipeline construction
	Other Obstacles	Other objects have obstructed the flow

Ductile Iron Wastewater Pipe Failure Modes and Mechanisms (Force Main] *B. Joint*		
Failure Mode		Failure Mechanism
Typical Failure Modes		
1	**Push-On**	
	Gasket/Seal Failure	Installation, stray currents, corrosive ground
	Angular Deflection	Installation, ground movement, poor bedding
	Axial Pull Out	Installation, internal water pressure, water hammer
	Thermal Expansion/Craked Spigot	Installation, high temperature, freezing-thawing
	Lateral Pipe Movement	Installation, ground movement, water hammer
	Internal Corrosion	Installation, exposure to corrosive substances
	External Corrosion	Installation, corrosive ground, stray currents
2	**Bolted Flange**	
	Bolt Failure	Installation, stray currents, corrosive ground
	Flange Failure	Installation, stray currents, corrosive ground
	Gasket Failure	Installation, stray currents, corrosive ground
	Stress at Flange	Installation, deflection of the pipe
	Bolt Corrosion	Installation, stray currents, corrosive ground
3	**Restrained Joint**	
	Pull Out	Installation, internal pressure, ground movement
	External Corrosion	Installation, stray currents, corrosive ground
Rare Failure Modes		
4	**Mechanical Joint**	
	Bolt Failure	Installation, stray currents, corrosive ground
	Bolt Corrosion	Installation, stray currents, corrosive ground
	Backing Ring Corrosion	Installation, stray currents, corrosive ground
	Structural Spigot Failure	Installation, stray currents
5	**Slip-On Coupler**	
	Internal Corrosion	Installation, expose to corrosive substances
	External Corrosion	Installation, stray currents, corrosive ground
	Bolt Corrosion	Installation, stray currents, corrosive ground
	Spigot Pull-Out	Installation, ground movement, water hammer
6	**Ball & Socket**	
	Structural Socket Failure	Installation and stray currents

Appendix B-11

FORCE MAIN CAST IRON (CI) PIPE

Cast Iron Wastewater Pipe Failure Modes and Mechanisms (Force Main) *A. Main Line*		
Failure Mode		**Mechanism**
Typical Failure Mode		
Structural Failure		
1	**Crack**	
	Circumferential Crack	Thermal contraction, poor support leading to movement, external loading
	Longitudinal Crack	Internal water pressure, crushing forces acting along the pipe
	Spiral Crack	Combination of bending stress and hoop stress from internal pressure
	Mixed Crack	Combination of bending stress and hoop stress from internal pressure
2	**Fracture**	
	Circumferential Fracture	Thermal contraction, poor support leading to movement, external loading
	Longitudinal Fracture	Internal water pressure, crushing forces acting along the pipe
	Spiral Fracture	Combination of bending stress and hoop stress from internal pressure
	Mixed Fracture	Combination of bending stress and hoop stress from internal pressure
3	**Hole**	
	Blowout Holes	Corrosion pitting, casting flaws
4	**Corrosion**	
	External Corrosion	Exposure to corrosive substances, soil corrosivity, stray current
	Internal Corrosion	Exposure to corrosive substances, wastewater characteristics
5	**Lining Failure**	
	Spall	Poor installation, manufacturing defects, maintenance damage, impact damage during installation
	Crack	Poor installation, manufacturing defects, maintenance damage, impact damage during installation
Operational and Maintenance Failure		
6	**Exfiltration**	Holes, cracks caused by internal pressure and/or external loading
Rare Failure Mode		
Structural Failure		
	Shearing	
	Bell Shearing	Bending forces and/or compressive forces pushing the spigot of the pipe into the bell of the next pipeline
7	**Burst Failure**	
	Burst Failure (Int.)	Internal pitting and graphitization corrosion weakening wall often combined with induced strains
	Burst Failure (Ext.)	External pitting and graphitization corrosion weakening wall often combined with induced strains
	Burst Failure (3rd party)	Third party impact damage
8	**Collapsed**	excessive loading, extremely high pressure, vacuum loading, third party damage
Operational and Maintenance Failure		
9	**Obstacles/Obstructions**	
	Object Intruding Through Wall	Third party objects inserting through the pipe wall after the pipeline construction
	Object Wedged in the Joint	Object is wedged in the joint of pipeline
	Object Through Connection	Other objects entering into the main pipe via the connection and obstructed at the pipe connection
	External Pipe or Cable	Third party objects inserting through the pipe wall after the pipeline construction
	Other Obstacles	Other objects have obstructed the flow

Cast Iron Wastewater Pipe Failure Modes and Mechanisms (Force Main) *B. Joint*		
	Failure Mode	**Failure Mechanism**
Typical Failure Modes		
1	**Mechanical Joint**	
	Bolt Failure	Installation, stray currents, corrosive ground
	Bolt Corrosion	Installation, stray currents, corrosive ground
	Backing Ring Corrosion	Installation, stray currents, corrosive ground
	Structural Spigot Failure	Installation, stray currents, corrosive ground
2	**Bolted Flange**	
	Bolt Failure	Installation, stray currents, corrosive ground
	Flange Failure	Installation, stray currents, corrosive ground
	Gasket Failure	Installation, stray currents, corrosive ground
	Stress at Flange	Installation, deflection of the pipe
	Bolt Corrosion	Installation, stray currents, corrosive ground
3	**Slip-On Coupler**	
	Internal Corrosion	Installation, expose to corrosive substances
	External Corrosion	Installation, stray currents, corrosive ground
	Bolt Corrosion	Installation, stray currents, corrosive ground
	Spigot Pull-Out	Installation, ground movement, water hammer
4	**Groove & Shoulder**	
	Structural Socket Failure	Installation, stray currents
5	**Ball & Socket**	
	Structural Socket Failure	Installation, stray currents
Rare Failure Modes		
6	**Push-On**	
	Gasket/Seal Failure	Installation, stray currents, corrosive ground
	Angular Deflection	Installation, ground movement, poor bedding
	Axial Pull Out	Installation, internal water pressure, water hammer
	Thermal Expansion/Cracked Spigot	Installation, high temperature, freezing-thawing
	Lateral Pipe Movement	Installation, ground movement, water hammer
	Internal Corrosion	Installation, exposure to corrosive substances
	External Corrosion	Installation, corrosive ground, stray currents
7	**Ball & Spigot Lead/Jute**	
	Angular Deflection	Installation, stray currents
	Axial Pull Out	Installation, stray currents
	Lead/Jute Deterioration	Installation, stray currents
	Thermal Expansion/Cracked Spigot	Installation, stray currents
	Lateral Pipe Movement	Installation, stray currents
	External Corrosion	Installation, stray currents

Appendix B-12

FORCE MAIN POLYVINYL CHLORIDE (PVC) PIPE

PVC Wastewater Pipe Failure Modes and Mechanisms (Force Main)

A. Main Line

Failure Mode		Mechanism
Typical Failure Mode		
Structural Failure		
1	**Crack**	
	Longitudinal crack	Internal water pressure, crushing forces acting along the pipe, bending
	Spiral crack	Combination of bending stress and hoop stress from internal pressure
	Bell Splitting	Over insertion, excessive axial alignment, water pressure
2	**Fracture**	
	Longitudinal fracture	Excessive crushing or ring stress from non-uniform bedding support differential settlement and ground movement
	Spiral fractures	Pressure surges and/or combination of bending forces and internal pressure
3	**Broken**	Occurs when pieces of a cracked or fractured pipe move form their original position
4	**Hole**	Broken pipe has completely dislodged from the wall
5	**Deformed**	
	Axial	External pressure, settlement, uplift, over insertion and interventions
6	**Leakage/Burst**	
	Burst	Internal pressure, external pressure, axial tension, axial compression, bending, traffic load, settlement, uplift, production, impact, vibration, fatigue, fire, frost, abrasion, chemicals, biological, and interventions
	Burst Main	Internal pressure, external pressure, axial tension, axial compression, bending, traffic load, settlement, uplift, production, impact, vibration, fatigue, fire, frost, abrasion, chemicals, biological, and interventions
	Burst Section	Internal pressure, external pressure, axial tension, axial compression, bending, traffic load, settlement, uplift, production, impact, vibration, fatigue, fire, frost, abrasion, chemicals, biological, and interventions
	Broken Section	Internal pressure, external pressure, axial tension, axial compression, bending, traffic load, settlement, uplift, production, impact, vibration, fatigue, fire, frost, abrasion, chemicals, biological, and interventions
	Blown Section	Propagating brittle fracture bifurcates and then rejoins the original crack path
7	**Color Change**	UV radiation, chemicals, biological, and excessive stresses
8	**Blisters/Voids**	
	Voids	Production and chemicals
Operational and Maintenance Failure		
9	**Exfiltration**	Holes, cracks caused by internal pressure and/or external loading
Rare Failure Mode		
Structural Failure		
10	**Crack**	
	Circumferential crack	Thermal contraction, poor support leading to movement
	Mixed Cracking	Production, impact, vibration, fatigue, residual stresses, UV radiation, frost, chemicals, biological, and interventions
	Irregular Cracks	Environmental and stress cracking, pipe surface embrittled by UV, Chemical attacks
	Rapid Crack Propagation (RCP)	Dynamic loading at low temperatures
11	**Fracture**	
	Circumferential fracture	Vertical movement a result of excessive shear and bending stress
	Mixed fractures	Combination of stresses that cause longitudinal and circumferential cracking
12	**Crazing**	
	Through Crazing	Internal pressure, axial tension, settlement, fatigue, chemicals, and residual stresses, bad manufacturing
	External Crazing	Internal pressure, axial tension, settlement, fatigue, chemicals, and residual stresses, bad manufacturing
	Internal Crazing	Internal pressure, axial tension, settlement, fatigue, chemicals, and residual stresses, bad manufacturing
13	**Deformed**	
	Ring	External pressure, axial tension, axial compression, bending, traffic load, settlement, uplift, high temperatures, fire, and interventions
14	**Buckling**	
	Axial Buckling	External pressure, axial compression, temperatures, fire, and interventions
	Transverse/Ring Buckling	External pressure, axial compression, production, residual stresses, high temperatures, fire, and interventions
	Non-Symmetric Buckling	Longitudinal bending and brazier effect
	Longitudinal Buckling	Axial compression and thermal effects
	Other	External pressure, axial compression, production, residual stresses, high temperatures, fire, and interventions
15	**Leakage/Burst**	
	Snap Through	External pressure and hard external bending
	Two-Sided Snap Through	External pressure
	Perforation	Stress intensity factor doesn't exceed the material fracture toughness before the crack transverses the full wall thickness
	Pinhole	External point load, impact, and third party damage, minor imperfection during manufacturing, internal diameter tears
16	**Dimension Change**	Internal pressure, high temperatures, chemicals, and fire
17	**Blisters/Voids**	
	External Blister	Excessive amount of heat
	Internal Blister	Production, chemicals

Failure Mode		Mechanism
	Blazes	Osmosis
	Delamination	Production, improper fusion (acid tone test)
18	**Abrasion/Erosion**	Flow velocity and interventions
19	**Chemical Attacks**	Expose to Chemical such as solvent or oil
20	**Collapse**	External loading, third party damage
21	**Other**	
	Blank Ends	attachment, not properly joint
	Blown T	attachment, not properly joint
	Fire Plug Riser Split	attachment, not properly joint
	Ball Valve	attachment, not properly joint
	End Blown Off	attachment, not properly joint
Operational and Maintenance Failure		
22	**Obstacles/Obstructions**	
	Object intruding through wall	Third party objects inserting through the pipe wall after the pipeline construction
	Object wedged in the joint	Object is wedged in the joint of pipeline
	Object through connection	Other objects entering into the main pipe via the connection and obstructed at the pipe connection
	External pipe or cable	Third party objects inserting through the pipe wall after the pipeline construction
	Other Obstacles	Other objects have obstructed the flow

PVC Wastewater Pipe Failure Modes and Mechanisms (Force Main)		
B. Joint		
	Failure Mode	Mechanism
Typical Failure Mode		
1	**Push-fit and slip-on collar (Gasket)**	
	Displaced gasket (into socket)	Poor installation, improper alignment
	Cracking of socket	Over insertion, axial deflecting joint
Rare Failure Modes		
2	**Push-fit and slip-on collar (Gasket)**	
	Spigot deflection	Deflection of the pipe, differential settlement
	Gasket degradation	Exposure to chemical (solvent, oil), UV
	Bell splitting	Surge fatigue, water hammer
	PVC material degradation	Exposure to chemical (solvent, oil), UV
3	**Solvent Weld**	
	Poor Adhesion degradation	Poor Adhesive
	Embrittlement	Temperature changes
	Bell splitting	Surge fatigue, water hammer
	Angular deflection	Ground movement, poor bedding
	Misalignment	Poor installation, ground movement
4	**Other**	
	Joint Leak	
	Mechanical Joint Leak	Improper Tightening
	Blown Rubber on Flange	
	Blown Joint	
	Leak on Collar	Gasket degradation
	Leaking Rubber Ring	Material degradation
	Gasket/Seal Failure	Displaced gasket into socket, gasket degradation
	Structural Bell Failure	Cracking of socket
	Butt Fusion Joint	Weak fusion weld under axial, shear, and bending forces
	Electro-Fusion Joint	Axial tension, bending, internal pressure, and improper welding
	Flange Joint	Improper welding, axial tension, and bending

Appendix B-13

FORCE MAIN STEEL (STL) PIPE

Steel Wastewater Pipe Failure Modes and Mechanisms (Force Main) *A. Main Line*		
Failure Mode		**Mechanism**
Typical Failure Mode		
Structural Failure		
1	**Hole**	
	Blowout Holes	Corrosion pitting
2	**Corrosion**	
	External Corrosion	Exposure to corrosive substances, soil corrosivity, stray current
	Internal Corrosion	Exposure to corrosive substances
3	**Deformed**	
	Vertical Deformed	pipe loses the support of the surrounding ground, overloading, impact loading
	Horizontal Deformed	pipe loses the support of the surrounding ground, overloading, impact loading
4	**Lining Failure**	
	Detached	Poor installation, manufacturing defects, maintenance damage, impact damage during installation
	Defective End	Poor installation, manufacturing defects, maintenance damage, impact damage during installation
	Blistered	Poor installation, manufacturing defects, maintenance damage, impact damage during installation
	Service Cut Shifted	Poor installation, manufacturing defects, maintenance damage, impact damage during installation
	Abandoned Connection	Poor installation
	Overcut Service	Poor installation, liners shrink after the services are cut
	Undercut Service	Poor installation
	Buckled	Poor installation, manufacturing defects, maintenance damage, impact damage during installation
	Wrinkled	Poor installation, manufacturing defects, maintenance damage, impact damage during installation
	Spall	Poor installation, manufacturing defects, maintenance damage, impact damage during installation
	Crack	Poor installation, manufacturing defects, maintenance damage, impact damage during installation
5	**Burst Failure**	
	Burst Failure (Int.)	Internal pitting and graphitization corrosion weakening wall often combined with induced strains
	Burst Failure (Ext.)	External pitting and graphitization corrosion weakening wall often combined with induced strains
	Burst Failure (3rd party)	Third party impact damage
6	**Weld Failure**	
	Longitudinal	Poor Welding, filling material used
	Circumferential	Poor Welding, filling material used
	Multiple	Poor Welding, filling material used
	Spiral	Poor Welding, filling material used
Operational and Maintenance Failure		
7	**Exfiltration**	Holes, cracks caused by internal pressure and/or external loading
Rare Failure Mode		
Structural Failure		
8	**Crack**	
	Circumferential Crack	Thermal contraction, poor support leading to movement, external loading
	Longitudinal Crack	Internal water pressure, crushing forces acting along the pipe
	Spiral Crack	Combination of bending stress and hoop stress from internal pressure
	Mixed Crack	Combination of bending stress and hoop stress from internal pressure
9	**Fracture**	
	Circumferential Fracture	Thermal contraction, poor support leading to movement, external loading
	Longitudinal Fracture	Internal water pressure, crushing forces acting along the pipe
	Spiral Fracture	Combination of bending stress and hoop stress from internal pressure
	Mixed Fracture	Combination of bending stress and hoop stress from internal pressure
	Shearing	
	Bell Shearing	Bending forces and/or compressive forces pushing the spigot of the pipe into the bell of the next pipeline
10	**Collapsed**	excessive loading, extremely high pressure, vacuum loading, third party damage
Operational and Maintenance Failure		
11	**Obstacles/Obstructions**	
	Object Intruding Through Wall	Third party objects inserting through the pipe wall after the pipeline construction
	Object Wedged in the Joint	Object is wedged in the joint of pipeline
	Object Through Connection	Other objects entering into the main pipe via the connection and obstructed at the pipe connection
	External Pipe or Cable	Third party objects inserting through the pipe wall after the pipeline construction
	Other Obstacles	Other objects have obstructed the flow

Steel Wastewater Pipe Failure Modes and Mechanisms (Force Main) *B. Joint*		
Failure Mode		Failure Mechanism
Typical Failure Modes		
1	**Push-On**	
	Gasket/Seal Failure	Installation, stray currents, corrosive ground
	Angular Deflection	Installation, ground movement, poor bedding
	Axial Pull Out	Installation, internal water pressure, water hammer
	Thermal Expansion/Craked Spigot	Installation, high temperature, freezing-thawing
	Lateral Pipe Movement	Installation, ground movement, water hammer
	Internal Corrosion	Installation, exposure to corrosive substances
	External Corrosion	Installation, corrosive ground, stray currents
2	**Bolted Flange**	
	Bolt Failure	Installation, stray currents, corrosive ground
	Flange Failure	Installation, stray currents, corrosive ground
	Gasket Failure	Installation, stray currents, corrosive ground
	Stress at Flange	Installation, deflection of the pipe
	Bolt Corrosion	Installation, stray currents, corrosive ground
Rare Failure Modes		
3	**Mechanical Joint**	
	Bolt Failure	Installation, stray currents, corrosive ground
	Bolt Corrosion	Installation, stray currents, corrosive ground
	Backing Ring Corrosion	Installation, stray currents, corrosive ground
	Structural Spigot Failure	Installation, stray currents
4	**Slip-On Coupler**	
	Internal Corrosion	Installation, expose to corrosive substances
	External Corrosion	Installation, stray currents, corrosive ground
	Bolt Corrosion	Installation, stray currents, corrosive ground
	Spigot Pull-Out	Installation, ground movement, water hammer

Appendix C

WASTEWATER PIPE DATA STRUCTURES

Appendix C-1

WOOD DATA STRUCTURE

(ESSENTIAL & PREFERABLE)

Wood Standard Wastewater Data Model (Essential Parameters)

No.	Parameter	Unit	Brief Explanation (See Attached Sheet for Detailed Explanation)	Confidence in Data (DR, DI, or EG)
	1	2	3	4
1	Node Identification Number	node	*ID for each pipe segments (Manhole-Manhole) between nodes*	
2	Pipe Material	type	*Different pipe material may deteriorate at different rate*	
3	Pipe Age	year	*Older pipes may deteriorate faster than newer pipes*	
4	Pipe Diameter	inch	*Different pipe sizes may fail in different failure modes*	
5	Function of Pipe	type	*Different use of sewer may deteriorate at different rates; Gravity, Combined, Force Main*	
6	Soil Type	type	*Some soils are corrosive, expansive, and compressible. Some soils contain hydrocarbons and solvents that may cause pipe deterioration.*	

* Confidence in Data: Direct Record (DR), Derived Indirectly (DI), or Educated Guess (EG)

Wood Standard Wastewater Data Model (Preferable Parameters)

No.	Parameter	Unit	Brief Explanation (See Attached Sheet for Detailed Explanation)	Confidence in Data (DR, DI, or EG)
	1	2	3	4
1	Pipe Depth	feet	*Pipe depth affects pipe loading and deteriorating rate*	
2	Pipe Joint Type	Type	*Some types of joints may undergo premature failure*	
3	Pipe Renewal Record	Record	*All records of pipes renewal- type of renewal method*	
4	Pipe Failure Record	record	*Record of Failure that occur, Failure mode should be specified, condition*	
5	Annual Capital Cost	dollar/year	*Annual capital cost of pipe*	

* Confidence in Data: Direct Record (DR), Derived Indirectly (DI), or Educated Guess (EG)

Appendix C-3

SILVER DATA STRUCTURE (ESSENTIAL & PREFERABLE)

Bronze Standard Wastewater Data Model (Essential Parameters)

No.	Parameter	Unit	Brief Explanation (See Attached Sheet for Detailed Explanation)	Confidence in data
	1	2	3	(*DR, DI, or EG)
Physical/Structural				
1	Node Identification Number	Node	*ID for each pipe segments (Manhole-Manhole) between nodes*	
2	Pipe Material	Type	*Different pipe materials deteriorate at different rates*	
3	Pipe Diameter	Inch	*Different pipe sizes may fall in different failure modes*	
4	Pipe Age	Year	*Older pipes may deteriorate faster than newer pipe*	
5	Pipe Depth	Feet	*Pipe Depth affects pipe loading and deteriorating rate*	
6	Pipe Joint Type	Type	*Some types of joints may undergo premature failure*	
7	Function of pipe	Type	*Different use of sewer may deteriorate at different rates; Combined, Sewer, Forced*	
Operational/Functional				
8	Pipe Renewal Record	Record	*All records of pipes renewal- type of renewal method*	
9	Pipe Failure Record	Record	*Record of Failure that occur, Failure mode should be specified*	
Environmental				
10	Soil Type	Type	*Corrosive, expansive, & compressible; hydrocarbons & solvents cause deterioration*	
Financial				
11	Annual Capital Cost	$/Year	*Utility annual capital Cost and allocation criteria*	

* Confidence in data: Direct Record (DR), Derived Indirectly (DI), or Educated Guess (EG)

Bronze Standard Wastewater Data Model (Preferable Parameters)

No.	Parameter	Unit	Brief Explanation (See Attached Sheet for Detailed Explanation)	Confidence in data
	1	2	3	(*DR, DI, or EG)
Physical/Structural				
1	Pipe Length	Feet	*Length of pipe between nodes (MH-MH)*	
2	Pipe Location	Area	*Some locations may receive roadway salt intrusion; urban, sub-urban, rural, costal, etc.*	
3	Pipe Shape	Type	*Different pipe shapes may result in different failure modes and deterioration*	
4	Pipe Bedding	Yes/No-Type	*Inadequate bedding may cause premature pipe failure, special bedding use*	
5	Trench Backfill	Type	*Some backfill materials are more corrosive or frost susceptible*	
6	Pipe Slope	Gradient	*Slope affects the velocity of gravity flow and may result in different pipe deterioration rates*	
7	Pipe Lining	Yes/No-Type	*Lined pipes have higher resistance to corrosion*	
Operational/Functional				
8	Operational & Maintenance Practices	Type- Level	*Poor practices can compromise structural integrity and water quality*	
9	Infiltration/Inflow	Level-Gal/Min	*infiltration/inflow may cause soil erosion, and increasing flow volume; Low,Med,High - also gal./min.*	
Environmental				
10	Loading Condition (Dead Load)	Lbs/sq.ft.	*Death load can be determined from infrastructure loading*	
11	Loading Condition (Live Load)	ADT-Level	*Live load can be determined from average daily traffic volume and railway loading etc.*	
12	Climate - Temperature	^{0}F	*Frost action in cold regions and seasonal soil water content variation in warmer regions*	

* Confidence in data: Direct Record (DR), Derived Indirectly (DI), or Educated Guess (EG)

Appendix C-3

SILVER DATA STRUCTURE (ESSENTIAL & PREFERABLE)

Silver Standard Wastewater Data Model (Essential Parameters)

No.	Parameter	Unit	Brief Explanation (See Attached Sheet for Detailed Explanation)	Confidence in data
	1	2	3	(*DR, DI, or EG)
Physical/Structural				
1	Node Identification Number	Node	*ID for each pipe segments (Manhole-Manhole) between nodes*	
2	Pipe Material	Type	*Different pipe materials deteriorate at different rates*	
3	Pipe Diameter	Inch	*Different pipe sizes may fall in different failure modes*	
4	Pipe Age	Year	*Older pipes may deteriorate faster than newer pipe*	
5	Pipe Depth	Feet	*Pipe Depth affects pipe loading and deteriorating rate*	
6	Node Length	Feet	*Length of pipe between nodes (MH-MH)*	
7	Pipe Location	Area	*Some locations may receive roadway salt intrusion; urban, sub-urban, rural, costal, etc.*	
8	Pipe Shape	Type	*Different pipe shapes may result in different failure modes and deterioration*	
9	Pipe Joint Type	Type	*Some types of joints may undergo premature failure*	
10	Function of pipe	Type	*Different use of sewer may deteriorate at different rates; Combined, Sewer, Forced*	
11	Pipe Bedding	Yes/No-Type	*Inadequate bedding may cause premature pipe failure, special bedding use*	
12	Trench Backfill	Type	*Some backfill materials are more corrosive or frost susceptible*	
13	Pipe Slope	Gradient	*Slope affects the velocity of gravity flow and may result in different pipe deterioration rates*	
14	Pipe Lining	Yes/No-Type	*Lined pipes have higher resistance to corrosion*	
Operational/Functional				
15	Operational & Maintenance Practices	Type- Level	*Poor practices can compromise structural integrity and water quality; very good, good, fair*	
16	Pipe Renewal Record	Record	*All records of pipes renewal - type of renewal method*	
17	Pipe Failure Record	Record	*Record of Failure that occur, Failure mode should be specified*	
18	Infiltration/Inflow	Level-Gal/Min	*infiltration/inflow may cause soil erosion, and increasing flow volume; Low,Med,High - also gal./min.*	
Environmental				
19	Soil Type	Type	*Corrosive, expansive, & compressible; hydrocarbons & solvents cause deterioration*	
20	Loading Condition (Dead Load)	Lbs/sq.ft.	*Death load can be determined from infrastructure loading*	
21	Loading Condition (Live Load)	ADT-Level	*Live load can be determined from average daily traffic volume and railway loading etc.*	
22	Climate - Temperature	°F	*Frost action in cold regions and seasonal soil water content variation in warmer regions*	
Financial				
23	Annual Capital Cost	$/Year	*Utility annual capital Cost and allocation criteria*	

* Confidence in data: Direct Record (DR), Derived Indirectly (DI), or Educated Guess (EG)

Silver Standard Wastewater Data Model (Preferable Parameters)

No.	Parameter	Unit	Brief Explanation (See Attached Sheet for Detailed Explanation)	Confidence in data
	1	2	3	(*DR, DI, or EG)
Physical/Structural				
1	Pipe Wall Thickness	Inch	*Wall thickness affects rupture resistance and corrosion penetration rates*	
2	Construction Specification	Spec.	*Construction specifications; Installation Circumstances*	
3	Design life of pipe	Year	*The pipe design life*	
4	Design strength of pipe	psi	*the pipe design strength (ring, longitudinal)*	
5	Manhole Condition	Record	*Manhole condition and relevant data*	
Operational/Functional				
6	Wastewater Quality	Record	*records of wastewater quality tested including PH, BOD, COD, and temperature*	
7	Wastewater Pressure	psi	*Internal water pressure affects pipe stresses and deterioration rate*	
8	Pipe Hydraulics	Gallon/Min	*Capacity of the sewage gravity conveying pipe*	
9	Pipe Surcharging	Yes/No - Ft.	*Surcharging in gravity sewers in dry & wet weather should be considered, head level in feet*	
10	Exfiltration	Level	*Exfiltration may cause erosion of soil and change soil loading on pipe; Low,Med,High*	
11	Blockage/stoppage	Yes/No-Type	*Blockage make the pipeline network inoperative, sewer pipe is no longer functional*	
12	Sediments	Ton/Feet	*Sediments per unit length*	
13	Inspection record	Yes/No-Record	*Record of inspection, method use, date of inspection*	
14	Flow Velocity	Feet/Second	*Low velocity accumulate deposits; excessive velocity accelerate deterioration at invert*	
Environmental				
15	Soil Corrosivity	Level	*Condition of the soil related to pipe deteriorate; low, medium, high*	
16	Soil Resistivity	Level	*Soils with low electrical resistivity are more likely to have high corrosion rates*	
17	Redox Potential	Level, mV	*Low Redox potentials are more favorable for sulphate reducing bacteria leading to corrosion*	
18	Soil Moisture Content	Percent	*Moisture percentage in the soil may affect loading and pipe deterioration*	
19	Stray Currents	Yes/No	*Stray currents may cause electrolytic corrosion of metal pipes*	
20	Groundwater Table	Feet	*affecting soil loading on the pipes and pipe deterioration rate; above, below sewer, fluctuating*	
21	Ground Cover	Type	*Paved ground or vegetation cover result in different deterioration mode and rate*	
22	Rainfall/Precipitation	Inch/year	*Rainfall in the areas should be monitored*	
23	Topography	Map	*Topography is very important for the performance of pipes; contour maps*	
24	Extreme Events	Yes/No - Type	*Information related to extreme events*	
Others				
25	Customer Complaint	Type	*Complaints related to blockage, flooding, pollutions, etc.*	
26	Chemical	Event	*Hydrogen sulphide may corrode pipe, etc.*	
27	FOG	Yes/No	*Fats, Oils, and Grease entering the sewer system*	
28	Overall Pipe Condition	Rating	*Condition of the pipe may be ranked from inspection tests; CCTV, smoke test, etc. (1-5)*	

* Confidence in data: Direct Record (DR), Derived Indirectly (DI), or Educated Guess (EG)

Appendix C-4

GOLD DATA STRUCTURE (ESSENTIAL & PREFERABLE)

Gold Standard Wastewater Data Model (Essential Parameters)

No.	Parameter	Unit	Brief Explanation (See Attached Sheet for Detailed Explanation)	Confidence in data
	1	2	3	(*DR, DI, or EG)
Physical/Structural				
1	Node Identification Number	Node	*ID for each pipe segments (Manhole-Manhole) between nodes*	
2	Pipe Material	Type	*Different pipe materials deteriorate at different rates*	
3	Pipe Diameter	Inch	*Different pipe sizes may fall in different failure modes*	
4	Pipe Age	Year	*Older pipes may deteriorate faster than newer pipe*	
5	Pipe Depth	Feet	*Pipe Depth affects pipe loading and deteriorating rate*	
6	Node Length	Feet	*Length of pipe between nodes (MH-MH)*	
7	Pipe Wall Thickness	Inch	*Wall thickness affects rupture resistance and corrosion penetration rates*	
8	Pipe Location	Area	*Some locations may receive roadway salt intrusion; urban, sub-urban, rural, costal, etc.*	
9	Pipe Shape	Type	*Different pipe shapes may result in different failure modes and deterioration*	
10	Pipe Joint Type	Type	*Some types of joints may undergo premature failure*	
11	Function of pipe	Type	*Different use of sewer may deteriorate at different rates; Combined, Sewer, Forced*	
12	Pipe Bedding	Yes/No-Type	*Inadequate bedding may cause premature pipe failure, special bedding use*	
13	Trench Backfill	Type	*Some backfill materials are more corrosive or frost susceptible*	
14	Construction Specification	Spec.	*Construction specifications; Installation Circumstances*	
15	Pipe Slope	Gradient	*Slope affects the velocity of gravity flow and may result in different pipe deterioration rates*	
16	Design life of pipe	Year	*The pipe design life in year*	
17	Design strength of pipe	psi	*the pipe design strength (ring, longitudinal)*	
18	Pipe Lining	Yes/No-Type	*Lined pipes have higher resistance to corrosion*	
19	Manhole Condition	Record	*Manhole condition and relevant data*	
Operational/Functional				
20	Wastewater Quality	Record	*records of wastewater quality tested including PH, BOD, COD, and temperature*	
21	Wastewater Pressure	psi	*Internal water pressure affects pipe stresses and deterioration rate*	
22	Pipe Hydraulics	Gallon/Min	*Capacity of the sewage gravity conveying pipe*	
23	Pipe Surcharging	Yes/No - Ft.	*Surcharging in gravity sewers in dry & wet weather should be considered, head level in feet*	
24	Operational & Maintenance Practices	Type- Level	*Poor practices can compromise structural integrity and water quality; very good, good, fair*	
25	Pipe Renewal Record	Record	*All records of pipes renewal- type of renewal method*	
26	Pipe Failure Record	Record	*Record of Failure that occur, Failure mode should be specified*	
27	Infiltration/Inflow	Level-Gal/Min	*infiltration/inflow may cause soil erosion, and increasing flow volume; Low,Med,High - also gal./min.*	
28	Exfiltration	Level	*Exfiltration may cause erosion of soil and change soil loading on pipe; Low,Med,High*	
29	Blockage/stoppage	Yes/No-Type	*Blockage make the pipeline network inoperative, sewer pipe is no longer functional*	
30	Sediments	Ton/Feet	*Sediments per unit length*	
31	Inspection record	Yes/No-Record	*Record of inspection, method use, date of inspection*	
32	Flow Velocity	Feet/Second	*Low velocity accumulate deposits; excessive velocity accelerate deterioration at invert*	
Environmental				
33	Soil Type	Type	*Corrosive, expansive, & compressible; hydrocarbons & solvents cause deterioration*	
34	Soil Corrosivity	Level	*Condition of the soil related to pipe deteriorate; low, medium, high*	
35	Soil Resistivity	Level	*Soils with low electrical resistivity are more likely to have high corrosion rates*	
36	Redox Potential	Level, mV	*Low Redox potentials are more favorable for sulphate reducing bacteria leading to corrosion*	
37	Soil Moisture Content	Percent	*Moisture percentage in the soil may affect loading and pipe deterioration*	
38	Stray Currents	Yes/No	*Stray currents may cause electrolytic corrosion of metal pipes*	
39	Groundwater Table	Feet	*affecting soil loading on the pipes and pipe deterioration rate; above, below sewer, fluctuating*	
40	Ground Cover	Type	*Paved ground or vegetation cover result in different deterioration mode and rate*	
41	Loading Condition (Dead Load)	Lbs/sq.ft.	*Death load can be determined from infrastructure loading*	
42	Loading Condition (Live Load)	ADT-Level	*Live load can be determined from average daily traffic volume and railway loading etc.*	
43	Rainfall/Precipitation	Inch/year	*Rainfall in the areas should be monitored*	
44	Climate - Temperature	ºF	*Frost action in cold regions and seasonal soil water content variation in warmer regions*	
45	Topography	Map	*Topography is very important for the performance of pipes; contour maps*	
46	Extreme Events	Yes/No - Type	*Information related to extreme events*	
Financial				
47	Annual Capital Cost	$/Year	*Utility annual capital Cost and allocation criteria*	
Others				
48	Customer Complaint	Type	*Complaints related to blockage, flooding, pollutions, etc.*	
49	Chemical	Event	*Hydrogen sulphide may corrode pipe, etc.*	
50	FOG	Yes/No	*Fats, Oils, and Grease entering the sewer system*	
51	Overall Pipe Condition	Rating	*Condition of the pipe may be ranked from inspection tests; CCTV, smoke test, etc. (1-5)*	

* Confidence in data: Direct Record (DR), Derived Indirectly (DI), or Educated Guess (EG)

Gold Standard Wastewater Data Model (Preferable Parameters)

No.	Parameter	Unit	Brief Explanation (See Attached Sheet for Detailed Explanation)	Confidence in data
	1	2	3	(*DR, DI, or EG)
Physical/Structural				
1	Pipe Quality	Level	*Imperfect pipes may deteriorate faster; poor, fair, good*	
2	Pipe Section Length	Feet	*Length of pipe section (Joint - joint)*	
3	Pipe Vintage	Year	*Pipes made at different time and place may deteriorate differently*	
4	Pipe Lateral	Type	*Some types and materials of laterals may undergo premature failure*	
5	Dissimilar Materials	Yes/No	*Dissimilar metals/materials are more susceptible to galvanic corrosion*	
6	Pipe Installation	Rating	*Improper Installation may cause pipe damage and increase deterioration rate*	
7	Pipe Manufacture	Record	*Defects in pipe walls produced by manufacturing errors can make pipes vulnerable to failure*	
8	Pipe Trench Width	Feet	*Trench width may affects soil loading on the pipes and deterioration rate*	
9	Pipe external Coating	Yes/No - type	*external coating prevents corrosion of the pipe*	
10	Pipe Cathodic Protection	Yes/No - type	*Technique used to control the corrosion of a metal surface*	
11	Pipe Thrust Restraint	Yes/No - Type	*Inadequate restraint may increase longitudinal pipe stresses*	
12	Lateral Connections	Record	*Condition of lateral connections and other related information such as type of connection*	
13	Pumping Station and WWTP	Record	*Location of the pumping stations and wastewater treatment plants*	
Operational/Functional				
14	Sewer Odors	Yes/No	*Solids build-ups, poor system hydraulics, flat grade, etc.*	
15	Sewer Flooding	Yes/No	*Flooding may change property of surrounding soil and loading on pipe*	
16	Sewer Overflow (SSO/CSO)	Yes/No	*Overflow may inundate surrounding soil and change loading on pipe*	
17	Backup floodings	Number	*Number of properties affected by flooding in Dry & Wet weather*	
18	Leakage allowance	%	*Percentage of Leakage allowance*	
19	Interruptions	Record	*Interruption of wastewater collection services*	
Environmental				
20	Soil Disturbance	Yes/No	*Disturbance of soil may cause damage or change soil support or loading to the pipe*	
21	Runoff Rate	Cu. Ft/Sec.	*Excess water flow which can be caused by rainfall, evaporation, snow melting, etc.*	
22	Non-Uniform Soil	Yes/No	*Non-uniform soil support in longitudinal axis may increase shear and bending stresses*	
23	Frost Penetration	Yes/No-depth	*Soil ever frozen around the pipe, depth of penetration in feet.*	
24	non-Uniform slope	Yes/No	*non-uniform slope may reduce the operating performance*	
25	Unstable Slope	Yes/No	*Pipes in unstable slope may be subjected to downslope creep displacement*	
26	Seismic Activity	Yes/No	*Seismic loading may lead to pipe rupture and pressure surge*	
27	Catchment Area (Sewershed)	Sq.Ft	*Extent of area receiving the wastewater feeding a part or the totality of sewer*	
28	Average Closeness to Trees	Feet	*Average distance between sewer and trees*	
29	Tidal Influences	Yes/No	*Sewer in Coaster area may be subjected to tidal influence affecting bedding of the pipe*	
30	Soil pH	pH	*Low pH (<4) means soil is acidic and likely to promote corrosion; high alkaline conditions (pH>8) can also lead to high corrosion*	
31	Soil Chloride	%	*Mortar coating usually creates a pH environment of >12.4. Low chloride levels in high pH(>11.5) environments can lead to serious corrosion*	
32	Soil Sulfate	%	*Accounts for microbial induced corrosion (MIC) and possible food source for sulfate reducting bacteria in anaerobic conditions under loose coatings*	
33	Soil Sulfide	%	*Sulfate reducing bacteria giving off sulfides which are excellent electrolytes*	
34	Pipe Connections	Type	*Type of connections - Residential, Industrial, Commercial, Agricultural, others*	
Financial				
35	Annual Maintenance Cost	$/Year	*Routine Cleaning, etc.; Method and Cost of Maintenance*	
36	Annual Repair/Rehabilitation Cost	$/Year	*Method and Cost of Preservation and Improvement like grouting, lining, etc.*	
37	Installed and Replacement Cost	$	*Original cost of installation and replacement cost*	
38	Annual Operational Cost	$/Year	*cost spent each year for operating and functioning sewer system*	
39	Annual Energy Cost	$/Year	*Cost of energy use in sewer system i.e. Forced Main, Pumping station etc.*	
40	Depreciated Value	%	*Depreciated value and method of calculation*	
41	Benefit/Cost	Record	*Benefit-Cost Analysis*	
Others				
42	Density of Connections	Number/ Mile	*Number of properties connected to the sewer per mile*	
43	Resident Population Served	Number	*Total population living in the area that is responsibility of the system*	
44	Failing Utilities	Yes/No	*Failing wastewater or water pipes in a close proximity of the system*	
45	Consequence/Risk	Level	*Consequence of failures: low, medium, high*	
46	Third Party Damage	Yes/No	*Information related to third party damage*	
47	Other Information	-	*Information relevant for pipe condition assessment and deterioration modeling*	

* Confidence in data: Direct Record (DR), Derived Indirectly (DI), or Educated Guess (EG)

Appendix C-5

Gold Essential and Preferable Parameters Detail Description

Gold Standard Wastewater Pipe Data Model (Essential Parameters)

Key: Number. **Parameter Name**-*Parameter explanation.*

- Description: (unit of parameter)

Physical/Structural

1. **Pipe Identification Number**-*Identification of each pipes segments (Manhole-Manhole)*
   - Section ID: (Node)
2. **Pipe Material**-. *Different pipe materials deteriorate at different rates*
   - Pipe material can be classified as following: (Type)

a)	Asbestos Cement	AC
b)	Brick	BR
c)	Cast Iron	CI
d)	Corrugated Metal Pipe	CMP
e)	Concrete Pipe (Non-reinforcement)	CP
f)	Concrete Segments (bolted)	CSB
g)	Concrete Segments (unbolted)	CSU
h)	Clay Tile (not vitrified clay)	CT
i)	Ductile Iron	DI
j)	Fiberglass reinforced pipe	FRP
k)	Glass reinforced cement	GRC
l)	Pitch fiber (Orangeburg)	OB
m)	Polyethylene	PE
n)	Polypropylene	PP
o)	Plastic / steel composite	PSC
p)	Polyvinyl Chloride	PVC
q)	Prestressed Concrete Cylinder Pipe	PCCP

r) Reinforced concrete pipe RCP

s) Reinforced plastic matrix (truss pipe) RPM

t) Steel pipe STL

u) Transite TTE

v) Vitrified clay pipe VCP

w) Wood WD

x) Not know XXX

y) Others ZZZ

3. **Pipe Diameter**-*Different pipe sizes may fail in different failure modes.*

   - diameter of pipe: (Inch)

4. **Pipe Age**-*Older pipes may deteriorate faster than newer pipes.*

   - Age of pipe: (Year)

5. **Pipe Depth**-*Pipe depth affects pipe loading and deteriorating rate.*

   - Distance from ground level to the crown of the pipe: (Feet)

6. **Node Length**-*Length of pipe between nodes (Manhole-Manhole)*

   - Length: (Feet)

7. **Pipe Wall Thickness**-*Wall thickness affects rupture resistance and corrosion penetration rate.*

   - Original thickness of pipe wall: (Inch)
   - If available, subsequent inspections: (Inch)

8. **Pipe Location**-*Geographical location may affect the performance of pipe.*

   - Geographical Location: (Area)

     a) Urban

     b) Sub-urban

     c) Rural

     d) Coastal

e) Industrial

g) Agricultural

9. **Pipe Shape**-*Different pipe shapes may result in different failure modes and deteriorations.*

- Shape of pipe: (Type)

    a) Circular

    b) Rectangular

    c) Oval

    d) Egg

    e) Others

10. **Pipe Joint Type**-*Some types of joints may undergo premature failure.*

- Type of pipe joint: (Type)

    a) Lead Yarn Joints

    b) Flanged Joints

    c) Mechanical Joints

    d) Welded Joints

    e) Others

11. **Function of pipe**- *Different use of sewer may deteriorate at different rates*

- Function of wastewater pipe: (Type)

    a) Force main (sewage main)

    b) Gravity Sewer

    c) Combined Sewer

    d) Others

12. **Pipe Bedding**-*Inadequate bedding may cause premature pipe failure.*

- special bedding and soil type If used special bedding, also provide the type: (Yes/No-Type)

  a) Clay

  b) Granular

  c) Variable Soils

  d) Concrete

  e) Other Known material

  f) Unknown material

- If available, Installation method: (Type)
- If available, Height of bedding: (Inch)

13. **Trench Backfill**-*Some backfill materials are more corrosive or frost susceptible.*

- Trench backfill soil: (Type)

  a) Clay

  b) Granular

  c) Variable Soils

  d) Other Known Soil

  e) Unknown Soil

14. **Construction Specification**-*Improper construction specifications may result in faulty pipe installation.*

- Construction specifications used: (Specification)

15. **Pipe Slope**-*Slope affects the velocity of gravity flow and may result in different pipe deterioration rates.*

- Slope: (Gradient)

16. **Design Life of Pipe**-*Original design life of each pipe.*

- Life: (Year)

17. **Design Strength of Pipe**-*Original design strength of each pipe.*

- Strength: (psi)

    a) Longitudinal Strength

    b) Ring Strength

18. **Pipe Lining** -*Lined pipes have higher resistance to corrosion and reduce infiltration.*

- Lining of pipe: (Yes/No - Type)

    a) Cured in place (CIP)

    b) Fold and Form or Deform/Reform (FF)

    c) Spiral Wound (SW)

    d) Segmented panel (SN)

    e) Segmented pipe (SP)

    f) Others (ZZZ)

19. **Manhole Condition**- *Condition of Manhole may affect pipe deterioration.*

- Manhole Condition: (Record)

Operational/Functional

20. **Wastewater Quality**- *records of wastewater quality tested including PH, BOD, Soluble BOD temperature, and oxygen level.*

- Wastewater Quality: (Record)

21. **Wastewater Pressure**-*Internal water pressure affects pipe stresses and deterioration rate.*

- Wastewater Pressure: (psi)

22. **Pipe Hydraulics**- Pipe Hydraulics may affect deteriorated rate

- Capacity of the sewage gravity conveying pipe: (Gallon/Min)

23. **Pipe Surcharging**-*Surcharging in gravity sewers in dry & wet weather should be considered.*

- Surcharging, if yes, should also provide Head level: (Yes/No-Ft.)

24. **Operational & Maintenance Practice**-*Poor Practices can compromise structural integrity and water quality.*

- specific practices: (Type)

  Mechanical

  a) Rodding

  b) Bucket Machine

  Hydraulic

  c) Balling

  d) Flushing

  e) Jetting

  f) Scooter

  g) Kites, Bags, and Poly pigs

  h) Silt Traps

  i) Grease traps and sand/oil interceptors

  j) Chemical

  k) Others

- Quality: (Level)

  a) Very Good

  b) Good

  c) Fair

25. **Pipe Renewal Record**-*All records of pipe repair/rehab /replace including method use*

- Renewal Record: (Record)

  a) Type: (Type)

  b) Others

26. **Pipe Failure Record**-*All records of pipe failures including failure modes.*

- Failure record: (Record)
- Type of Failures or Defects

  Structural

  a) Crack (C)
  b) Fracture (F)
  c) Broken (B)
  d) Hole (H)
  e) Deformed (D)
  f) Collapsed (X)
  g) Joint (J)
  h) Surface Damage (S)
  i) Lining Failure (LF)
  j) Weld Failure (WF)
  k) Point Repair (RP)
  l) Brickwork

  Operational and Maintenance

  a) Deposits (D)
  b) Roots (R)
  c) Infiltration (I)
  d) Obstacles/Obstructions (OB)
  e) Vermin (V)

- Dimension of Defects (e.g., Length, Width, and Depth)
- Orientation of Defects (e.g., circumferential, axial, or diagonal)

- Defects location

    a) Internal, external, or mid-wall

    b) angular location (i.e., 12, 3, 6, 9 O'clock)

Note: More detailed information on how to record Defects can be found in NASSCO's PACP manual

27. **Infiltration/inflow**-*Infiltration may cause soil erosion, and increasing flow volume.*

- Level of infiltration/inflow, if available, also provide in Gal/Min: (Level-Gallon/Minute)

    a) low

    b) Medium

    c) High

28. Exfiltration- *Exfiltration may cause soil erosion, and change soil loading on pipe.*

- Level of Exfiltration: (Level)

    a) low

    b) Medium

    c) High

29. **Blockage/Stoppage**-*Blockage/stoppage make the pipeline network inoperative, sewer pipe is no longer functional.*

- Blockage/Stoppage: (Yes/No- Type)

30. **Sediments**-*Sediments per unit length.*

- Amount of Sediment per feet: (Ton/feet)

31. **Inspection Record**-*Record of inspection i.e. by CCTV, Smoke test, Dye test*

- Record of inspection, date of inspection: (Yes/No-Record)

- Method Use: (Type)

    a) Visual Inspection

    b) Camera Inspection

    c) Closed Circuit Television (CCTV)

    d) Lamping Inspection

e) Dye Test

f) Smoke Test

g) Others

32. **Flow Velocity**-*Flow velocity may affect internal corrosion of unlined/coated pipes.*

- Velocity: (Feet/Second)

Environmental

33. **Soil Type**-*Some soils are corrosive, expansive, and compressible. Some soils contain hydrocarbons and solvents that may cause pipe deterioration.*

- Type of soil: (Type)

a) Clay

b) Granular

c) Mucks

d) Mud

e) Organic Soil

f) Variable Soils

g) Other Known Soil

h) Unknown Soil

34. **Soil Corrosivity**-*Soil present may be corrosive and may affect pipe environment.*

- Corrosivity level of soil: (Level)

a) Low

b) Medium

c) High

35. **Soil Resistivity**- *Soils with low electrical resistivity are more likely to have high corrosion rates.*

- Soil Resistivity: (Level, mV)

36. **Redox Potential**- *Low Redox potentials are more favorable for sulphate reducing bacteria leading to corrosion.*

- Redox Potential: (Level, mV)

37. **Soil Moisture Content**-*Moisture present in the soil may affect loading and pipe deterioration rate.*

- Moisture Content of the Soil: (Percent)

38. **Stray Currents**-*Stray electrical currents may cause electrolytic corrosion.*

- Electrical currents present near pipe: (Yes/No)

39. **Groundwater Table**-*Groundwater affects soil loading on the pipe and pipe deterioration rate.*

- Depth of water table: (Feet)

40. **Ground Cover**-*Paved ground or vegetation covers result in different deterioration mode and rate.*

- Land Cover: (Type)

    a) Grass

    b) Asphalt

    c) Concrete

    d) Trees

    e) Bare Ground

    f) Water

    g) Other Known Type

    h) Unknown Type

41. **Loading Condition (Dead Load)** -*loading depends on depth of pipe and infrastructure loading.*

- Death Load i.e. soil, structure, or stockpile above pipe: (Lbs/Sq.ft.)

42. **Loading Condition (Live Load)** - *Live load can be determined from average daily traffic volume and railway loading etc.*

- Live Load i.e. Traffic, Railway, Aircraft: (Average daily traffic volume (ADT) or Level)

    a) High Traffic

b) Medium Traffic

c) Low Traffic

43. **Rainfall/Precipitation**-*Rainfall in the areas should be monitored.*

- Rate of rainfall per year: (Inch/Year)

44. Climate-Temperature-Frost action in cold *regions may accelerate pipe deterioration.*

- Region temperature: (°F)

a) Average

b) High

c) Low

45. Topography-*Topography is very important for pipe performance.*

- Contour Maps illustrate topography of the land: (Map)

46. Extreme Events-*Events may threaten pipe sustainability.*

- Events: (Yes/No -Type)

a) Hurricanes

b) Floods

c) Tornadoes

d) Tsunami

e) Earthquakes

f) Other Known Event

g) Unknown Event

Financial

47. **Annual Capital Cost**- *Annual capital cost and allocation criteria.*

- Cost: ($/Year)

Others

48. **Customer Complaints**-*Complaints related to various problems.*

    - Complaints: (Type)

        a) Blockage complaints

        b) Flooding complaints

        c) Pollution complaints

        d) Odor complaints

49. **Chemistry**-*Some chemical substance may deteriorate pipe i.e. Hydrogen Sulphide may corrode pipe.*

    - Evidence of pipe exposed to chemical: (Event)

50. **FOG**-*Fats, Oils, and Grease entering the sewer system.*

    - FOG released to the system: (Yes/No)

51. **Overall Pipe Condition**-*Condition of the pipe may be ranked from inspection tests.*

    - Condition i.e. 1-5: (Rating)

Gold Standard Wastewater Pipe Data Model (Preferable Parameters)

Key: Number. **Parameter Name**-*Parameter explanation.*

- Description: (unit of parameter)

Physical/Structural

1. **Pipe Quality**-*Imperfect pipes may deteriorate faster.*
   - Quality i.e. poor, fair, good : (Level)
2. **Pipe Section Length**-*Length of pipe section (joint-joint).*
   - Length: (Feet)
3. **Pipe Vintage**-*Pipes made at a different time and place may deteriorate differently.*
   - When the pipe was made: (Year)

     a) After 1995

     b) 1985 - 1994 (inclusive)

     c) 1975 - 1984 (inclusive)

     d) 1950 - 1974 (inclusive)

     e) 1925 - 1949 (inclusive)

     f) Before 1925

     g) Unknown age
4. **Pipe Lateral**- *Some types of laterals may undergo premature failure.*
   - Lateral Material: (Type)

     a) PVC

     b) Concrete

     c) Cast-Iron

     d) Others
5. **Dissimilar Materials**-*Dissimilar materials are more susceptible to galvanic corrosion.*
   - Pipe made of different metals/materials: (Yes/No)

6. **Pipe Installation**-*Improper installation may cause pipe damage and increase deterioration.*

    - Installation method: (Rating)

        a) Proper

        b) Adequate

        c) Improper

7. **Pipe Manufacture**-*Defects in pipe walls produced by manufacturing errors can make pipes vulnerable to failure. This problem is most common in older pit cast pipes.*

    - Pipe Manufacture Record: (Record)
    - Should include pipe manufacturer, pipe class (e.g., AWWA 301, etc), and manufacture date.

8. **Pipe Trench Width**-*Trench width may affect soil loading on the pipes and deterioration rate.*

    - Trench Width: (Feet)

9. **Pipe External Coating** –*external coating prevents corrosion to the external wall.*

    - Coating: (Yes/No - Type)

10. **Pipe Cathodic Protection**-*Technique used to control the corrosion of a metal surface.*

    - Effective methods used to prevent stress corrosion cracking: (yes/no) If yes, type.

11. **Pipe Thrust Restraint**-*Inadequate restraint may increase longitudinal pipe stresses.*

    - Restraint present, holding or cradling the pipe, if yes, provide type: (Yes/No-Type)

        a) Thrust Block

        b) Restraint Joint

        c) Others

12. **Lateral Connections**-*Condition of laterals and other related information.*

    - Condition and other information: (Record)

13. **Pumping Station and WWTP**- *Location of the pumping stations and wastewater treatment plants.*

    - Pumping Station and WWTP: (Record)

Operational/Functional

14. **Sewer Odors**-*Solids build-ups, poor system hydraulics, flat grade, etc.*

    - Odors reported: (Yes/No)

15. **Sewer Flooding**-*Flooding may change property of surrounding soil and loading on pipe.*

    - Flooding : (Yes/No)

16. **Sewer Overflow (SSO/CSO)**-*Overflow may inundate surrounding soil and change loading on pipe.*

    - Overflow: (Yes/No)

17. **Backup Flooding**-*Number of properties affected by flooding in Dry & Wet weather.*

    - Number of Properties: (Number)

18. **Leakage Allowance**-*Percentage of leakage allowance.*

    - Leakage Allowance: (Percent)

19. **Interruptions**-*Interruption of wastewater collection services i.e. for maintenance or because of blockage*

    - Interruptions: (Yes/No)

Environmental

20. **Soil Disturbance**-*Disturbance of soil near the pipe may cause pipe damage or change soil support or loading.*

    - Any reason for soil disturbance around the pipe i.e. new construction: (Yes/No)

21. **Runoff Rate**-*Excess water flow which can be caused by rainfall, evaporation, snow melting, etc.*

    - Peak Runoff Rate: (Cubic feet/Second)

22. **Non-Uniform Soil**-*Non-uniform soil support in longitudinal axis may increase shear and bending stresses.*

- Non-Uniform Soil: (Yes/No)

23. **Frost Penetration**-*Pipe dwindling or defects may result from frozen soil.*

- Soil frozen near or around pipe, depth of penetration in feet: (Yes/No-Ft.)

24. **Non-Uniform Slope**-*Non-uniform slope may reduce operating performance.*

- Non- uniform Slope: (Yes/No)

25. **Unstable Slope**-*Pipes in unstable slope may be subjected to down slope creep displacement.*

- Slope of land above pipe unstable: (Yes/No)

26. **Seismic Activity**-*Seismic loading may lead to pipe rupture and pressure surge.*

- Seismic activity presents: (Yes/No)

27. **Catchment Area (Sewershed)**-*Extent of area receiving the wastewater feeding a part or the totality of sewer.*

- Area of drainage area: (Square Feet)

28. **Average Closeness to Trees**-*Root intrusion may cause pipe distress and accelerate pipe deterioration.*

- Average distance between sewer and trees: (Feet)

29. **Tidal Influences**-*Coastal areas with tidal influences may affect pipe bedding.*

- Tidal influences present: (Yes/No)

30. **Soil pH**-low or high *soil pH may accelerate the rate of deterioration.*

- pH: (pH)

31. **Soil Chloride**-*Mortar coating usually creates a pH environment of >12.4. Low chloride levels in high pH(>11.5) environments can lead to serious corrosion.*

- Chloride: (Percent)

32. **Soil Sulfate**-*Accounts for microbial induced corrosion (MIC) and possible food source for sulfate reducing bacteria in anaerobic conditions under loose coatings.*

- Sulfate: (Percent)

33. **Soil Sulfide**-*Accounts for microbial induced corrosion (MIC) and possible food source for sulfate reducing bacteria in anaerobic conditions under loose coatings.*

- Sulfide: (Percent)

34. **Pipe Connections**- *Type of Service connection may affect deteriorating rate*

- Type of connections: (Type)

  a) Domestic

  b) Industrial

  c) Commercial

  d) Agricultural

  e) Others

Financial

35. **Annual Maintenance Cost**- *Cost of Maintenance like Routine Cleaning etc.*

- Cost: (Dollar/Year)

36. **Annual Repair/Rehabilitation Cost**- *Cost of Preservation and Improvement like grouting, lining, etc.*

- Cost: (Dollar/Year)
- Expected Life of Repaired/ Rehabbed Pipe: (Year)

37. **Installed and Replacement Cost**-*Original cost of installation and replacement cost.*

- Cost: ($)

38. **Annual Operational Cost**- *Cost spent each year for operating and functioning sewer system*

- Cost: (Dollar/Year)

39. **Annual Energy Cost**-*Cost of energy use in sewer system i.e. Forced Main, Pumping station etc.*

- Cost: (Dollar/Year)

40. **Depreciated Value**-*Depreciated value and method of calculation.*

- Depreciated value: (%)
- Calculation Method

41. **Benefit/Cost**-*Benefit-Cost Analysis.*

- Benefit-Cost Analysis: (Record)

Others

42. **Density of Connections**-*Number of properties connected to the sewer per mile.*

- properties: (Number/mile)

43. **Resident Population Served**-*Total population living in the area that is responsibility of the system.*

- Number of population in the area: (Number)

44. **Failing Utilities**-*Failing water or wastewater pipes in a close proximity of the system.*

- Evidence: (Yes/No)

45. **Consequence/Risk**-*Consequence of the Failure.*

- Level of Consequence: (Level)

    a. Low

    b. Medium

    c. High

46. **Third Party Damage**-*Damages to pipe or bedding may be due to third parties.*

- Damages due to third party: (Yes/No)

47. **Other Information** –*Other related information that may affect pipe deterioration.*

Appendix D

WASTEWATER PIPE DATA COLLECTION METHODS AND PROTOCOLS

Appendix D-1

GOLD ESSENTIAL PARAMETERS DATA COLLECTION METHODS AND PROTOCOLS

Wastewater Pipeline Standard Data Collection Methods and Protocols

(Essential Parameters)

1. **Node Identification Number**

Sources: GIS, other databases

2. **Pipe Material**

Sources: GIS, as-built drawing, specifications, previous inspection records

3. **Pipe Nominal Diameter**

Sources: GIS, as-built drawing, specifications, previous inspection records, class of pipe

4. **Pipe Age**

Sources: GIS, as-built drawing, specifications, previous inspection records

5. **Pipe Depth**

Sources: Manhole Surveys, GIS, as-built drawings, specifications

6. **Node Length**

Sources: GIS, as-built drawing, specifications, previous inspection records

7. **Pipe Wall Thickness**

Sources: GIS, as-built drawing, specifications, class of pipe

Inspection/Testing:

- Ultrasonic measurement - measuring corrosion depths and total wall thickness.
- Magnetic permeability scanning - utilizing magnetic flux pigs propelled along the pipeline.
- Taking pipe sample

Ref:

- Nondestructive Testing Handbook, Birks, A. and Green, R., 1991
- Diagnostic Techniques for Sewer Systems, Makar, J., 1999
- Management and Maintenance Practices of Storm and Sanitary Sewers in Canadian Municipalities, Allouche, E., Freure, P., 2002
- Inspection Guidelines for Ferrous and Non-Ferrous Force Mains, WERF, Jason Consultants LLC, 2007

- A Multi-Sensory Approach to Structural Health Monitoring of Buried Sewer Pipelines Infrastructure System, Sinha, S. K., 2004
- Evaluation of Opportunities to Improve Structural Integrity Monitoring (SIM) Capabilities for Water Mains through Federal Research and Technology Transfer, EPA, 2006
- Protocols for Assessing Condition and Performance of Water and Wastewater Assets, WERF Condition Assessment Protocols (CAP) Project, Urquhart, T.,2006

8. Pipe Location

Sources: Manhole survey, GIS, as-build drawing, specifications, inspection records

Inspection/Testing:

- Radio Frequency Transponder
- Magnetics
- Electromagnetic
- Acoustic reflectometry and ground probing radar
- Test pit

Ref:

- Manual Manhole Inspection and Rehabilitation, American Society of Civil Engineers (ASCE), Engineering Practice No. 92
- Inspection Guidelines for Ferrous and Non-Ferrous Force Mains, WERF, Jason Consultants LLC, 2007

9. Pipe Shape

Sources: GIS, as-built drawing, specifications, previous inspection records

10. Pipe Joint Type

Sources: GIS, as-built drawing, specifications, previous inspection records

11. Function of pipe

Sources: GIS, as-built drawing, specifications, previous inspection records, operation and maintenance records

12. Pipe Bedding

Sources: As-built drawing, specifications, inspection records

13. Trench Backfill

Sources: As-built drawing, specifications

14. Construction Specification

Sources: Utility records

15. Pipe Slope

Sources: Manhole Surveys, GIS, as-built drawings, specifications

16. Design life of pipe

Sources: Pipe product standards, pipeline design codes, pipeline construction codes

17. Design Strength of Pipe

Sources: Pipe product standards, pipeline design codes, pipeline construction codes, Specifications

18. Pipe Lining

Sources: Repair and rehabilitation record, inspection records, construction specifications

19. Manhole Condition

Sources: Inspection records

Inspection/Testing:

- Vacuum test
- Air test

Ref:

- Manual Manhole Inspection and Rehabilitation, American Society of Civil Engineers (ASCE), Engineering Practice No. 92, USA
- Wastewater Collection Systems Management, Water Environment Federation (WEF), 1999

20. Wastewater Quality

Sources: Utility Records, Quality Control Reports

Inspection/Testing:

- Wastewater Temperature
- BOD Evaluation
- Wastewater pH
- Dissolved Oxygen, Dissolved Sulfide

Ref:

- Sewer System Infrastructure Analysis and Rehabilitation handbook, EPA, 1991
- Sulfide and Corrosion prediction and Control, American Concrete Pipe Association, 1998

21. Wastewater Pressure

Sources: Specifications, Utility Records

22. Pipe Hydraulics

Sources: Existing hydraulic model, flow monitoring log, construction specifications

Inspection/Testing:

- Flow Measurement
- Flow Monitoring

Ref:

- Wastewater Collection System Modeling and Design, First Edition, Heastad M. et al. , 2004
- Sewer System Infrastructure Analysis and Rehabilitation handbook, EPA, 1991
- Existing Sewer Evaluation and Rehabilitation, ASCE/WEF, 1994

23. Pipe Surcharging

Sources: Existing hydraulic model, Inspection records, operation and maintenance records

Inspection/Testing:

- Depth sensors
- Ultrasonic devices

Ref:

- Wastewater Collection System Modeling and Design, Heastad M. et al., 2004

24. Operational & Maintenance Practices

Sources: Operation and maintenance records, operations and maintenance manuals and reports

25. Pipe Renewal Record

Sources: Utility records

26. Pipe Failure Record

Sources: Utility records

27. Infiltration/Inflow

Sources: Inspection records, hydraulic model, flow monitoring, wastewater treatment works records, operation records

Inspection/Testing:

- Real-time continuous flow measurements - wastewater treatment plant as well as other permanent measuring points in the sewer system are used to monitoring Inflow & Infiltration.
- Temporary continuous flow monitoring- use of area velocity-flow meters.
- Pumping station monitoring – use of an event recorder logging on-off cycles of pumps.
- Inspection techniques - Smoke testing, Rainfall simulation (dye flooding and tracing), Building plumbing inspection, Manhole inspection, Flow isolation, TV inspection, Lateral testing
- Smoke Test
- Dye Test

Ref:

- Existing Sewer Evaluation and Rehabilitation, ASCE/WEF, 1994
- Infiltration/Inflow Control/Reduction for Wastewater Collection Systems, NRC CNRC Canada, 2003
- Sewer System Infrastructure Analysis and Rehabilitation, US EPA, 1991.
- Wastewater Collection Systems Management, Water Environment Federation (WEF) , 1999

28. Exfiltration

Sources: Inspection records, hydraulic model, flow monitoring, wastewater treatment works records, operation records, stream monitoring

Inspection/Testing:

- Real-time continuous flow measurements - wastewater treatment plant as well as other permanent measuring points in the sewer system are used to monitoring Inflow & Infiltration.
- Temporary continuous flow monitoring- use of area velocity-flow meters.
- Pumping station monitoring – use of an event recorder logging on-off cycles of pumps.
- Inspection techniques - Smoke testing, Rainfall simulation (dye flooding and tracing), Building plumbing inspection, Manhole inspection, Flow isolation, TV inspection, Lateral testing
- Smoke Test
- Dye Test

Ref:

- Existing Sewer Evaluation and Rehabilitation, ASCE/WEF, 1994
- Infiltration/Inflow Control/Reduction for Wastewater Collection Systems, NRC CNRC Canada, 2003

- Sewer System Infrastructure Analysis and Rehabilitation, US EPA, 1991.
- Wastewater Collection Systems Management, Water Environment Federation (WEF) , 1999

29. Blockage/Stoppage

Sources: Operation and maintenance records, Inspection records, customer service records

Inspection/Testing:

- Pipe cleaning
- CCTV inspections

Ref:

- Pipeline Assessment and Certification Program (PACP) Reference Manual, NASSCO, 2003
- Collection Systems O&M Fact Sheet Sewer Cleaning and Inspection, EPA, 1999

30. Sediments

Sources: Operation and maintenance records, inspections records

Inspection/Testing:

- Pipe cleaning
- CCTV inspections

Ref:

- Pipeline Assessment and Certification Program (PACP) Reference Manual, NASSCO, 2003
- Collection Systems O&M Fact Sheet Sewer Cleaning and Inspection, EPA, 1999

31. Inspection record

Sources: Utility records

32. Flow Velocity

Sources: Hydraulic model, permanent sewer flow monitors, wastewater treatment works records

Inspection/Testing:

- Flow meter

Ref:

- Wastewater Collection System Modeling and Design, Heastad M. et al., 2004
- Sulfide and Corrosion prediction and Control, American Concrete Pipe Association, 1998

33. Soil Type

Sources: GIS, other databases, Inspection records, field inspections, geotechnical records

- Geospatial data gateway - http://datagateway.nrcs.usda.gov/
- Natural Resources Conservation Service - http://soils.usda.gov/survey/geography/statsgo/

Inspection/Testing:

- American Society for Testing and Materials: ASTM D422 - 63(2007) Standard Test Method for Particle-Size Analysis of Soils
- American Society for Testing and Materials: ASTM D3282-93(2004)e1 Standard Practice for Classification of Soils and Soil-Aggregate Mixtures for Highway Construction Purposes
- American Association of State and Highway Transportation Officials : AASHTO M 145 Standard Specifications for Classification of Soils and Soil-Aggregate Mixtures for Highway Construction Purposes

34. Soil Corrosivity

Inspection/Testing:

- The 10-Point Soil Evaluation System in Appendix A of ANSI/AWWA C105/A21.5
- Metalogic Method
- Spickelmire Method
- Linear Polarization Resistance (LPR)

Ref:

- Protocols for Assessing Condition and Performance of Water and Wastewater Assets, WERF Condition Assessment Protocols (CAP) Project, Urquhart, T.,2006
- Fuzzy Expert System to Assess Corrosion of Cast/Ductile Iron Pipes from Backfill Properties, Najjaran H., 2006
- American National Standard for Polyethylene Encasement for Ductile-Iron, AWWA, 1999

35. Soil Resistivity

Inspection/Testing:

- ASTM G187 – 1997: Determination of the electrical resistivity of soil
- ASTM G187-05 Standard Test Method for Measurement of Soil Resistivity Using Two-Electrode Soil Box Method
- ASTM G57-95a (2001) Standard Test Method for Field Measurement of Soil Resistivity Using Wenner Four-Electrode Method

Ref:

- Protocols for Assessing Condition and Performance of Water and Wastewater Assets, WERF Condition Assessment Protocols (CAP) Project, Urquhart, T.,2006

36. Redox Potential

Inspection/Testing:

- Soil Characterization (Redox)

Ref:

- Protocols for Assessing Condition and Performance of Water and Wastewater Assets, WERF Condition Assessment Protocols (CAP) Project, Urquhart, T.,2006

37. Soil Moisture Content

Sources: Field inspections, geotechnical records

Inspection/Testing:

- American Society for Testing and Materials: ASTM D1558 - 99(2004)e1 Standard Test Method for Moisture Content Penetration Resistance Relationships of Fine-Grained Soils
- American Association of State and Highway Transportation Officials: AASHTO T-272-04 Family of Curves – One-Point Method

38. Stray Currents

Sources: GIS, other databases

Inspection/Testing:

- Pipe Potential Surveys

Ref:

- Stray Current Effects on Ductile Iron Pipe, Ductile Iron Pipe Research Association (DIPRA), Bonds R., 1997
- Protocols for Assessing Condition and Performance of Water and Wastewater Assets, WERF Condition Assessment Protocols (CAP) Project, Urquhart, T.,2006

39. Groundwater Table

Sources: Hydraulic Model

- U.S. Geological Survey (USGS) Ground-Water Data for the Nation - http://waterdata.usgs.gov/nwis/gw

40. Ground Cover

Sources: Arial photographs

- Geospatial data gateway - http://datagateway.nrcs.usda.gov/
- The Global Land Cover Facility - http://glcf.umiacs.umd.edu/data/landcover/data.shtml

41. Loading Condition (Dead Load)

Sources: Derived from imagery

Inspection/Testing:

- Loading Factor

Ref:

- Concrete Pipe Design Manual, American Concrete Pipe Association, 2000
- Concrete Pipe Handbook, American Concrete Pipe Association, 1998
- Design of Ductile Iron, Pipe Ductile Iron Pipe Research Association, 2003
- Clay Pipe Handbook, Clay Pipe Institute, 2004
- Load Factors and Non-Destructive Testing of Clay Pipe, Sikora J. E., 1980

42. Loading Condition (Live Load)

Sources: Derived from traffic information, Local department of transportation

Inspection/Testing:

- Loading Factor

Ref:

- Concrete Pipe Design Manual, American Concrete Pipe Association, 2000
- Concrete Pipe Handbook, American Concrete Pipe Association, 1998
- Design of Ductile Iron Pipe, Ductile Iron Pipe Research Association, 2003
- Clay Pipe Handbook, Clay Pipe Institute, 2004
- Load Factors and Non-Destructive Testing of Clay Pipe, Sikora J. E., 1980

43. Rainfall/Precipitation

Sources:

- Weather Warehouse - http://weather-warehouse.com,
- Weather Underground- http://www.wunderground.com,
- National Oceanic and Atmospheric Administration's National Weather Service - http://www.weather.gov,
- National Climate Data Center, US Department of Commerce - http://www.ncdc.noaa.gov/oa/ncdc.html
- Community Collaborative Rain, Hail and Snow Network - http://www.cocorahs.org/

Inspection/Testing:

- Rain gauge - measure amount of rainfall and its intensity. Rainfall is usually reported by daily, monthly, or yearly

44. Climate - Temperature

Sources:

- Weather Warehouse - http://weather-warehouse.com,
- Weather Underground- http://www.wunderground.com,
- National Oceanic and Atmospheric Administration's National Weather Service - http://www.weather.gov,
- National Climate Data Center, US Department of Commerce - http://www.ncdc.noaa.gov/oa/ncdc.html
- Community Collaborative Rain, Hail and Snow Network - http://www.cocorahs.org/

45. Topography

Sources:

- NM Bureau of Geology and Mineral Resources - http://geoinfo.nmt.edu/publications/maps/topographic/home.html
- TerraServer.com - http://www.terraserver.com/
- TopoZone.com - http://www.topozone.com
- USGS Earth Resources Observation and Science (EROS) - http://edc.usgs.gov

46. Extreme Events

Sources: Flooding records, national incident records, utility records

- national weather service - www.nws.noaa.gov/

47. Annual Capital Cost

Sources: Utility financial records

48. Customer Complaint

Sources: Utility records

49. Chemistry

Sources: Pollution incident records, industrial discharge licensing records

50. FOG

Sources: Utility records, pretreatment regulations

Inspection/Testing:

- Actual Oil and Grease build up in percentage

Ref:

- Sewer System Infrastructure Analysis and Rehabilitation handbook, EPA, 1991

51. Overall Pipe Condition

Sources: GIS, inspection records, operation and maintenance records

Inspection/Testing:

- Structural Evaluation
- CCTV inspection
- leakage detection

Ref:

- Inspection Guidelines for Ferrous and Non-Ferrous Force Mains, WERF, Jason Consultants LLC, 2007
- Pipeline Assessment and Certification Program (PACP), NASSCO, 2003

Appendix D-2

Gold Preferable Parameters Data Collection Methods and Protocols

Wastewater Pipeline Standard Data Collection Methods and Protocols

(Preferable Parameters)

1. Pipe Quality

Sources: Manufacturing logs, quality control reports, inspection records

2. Pipe Section Length

Sources: GIS, as-built drawings, construction specifications, inspection records, other database

3. Pipe Vintage

Sources: GIS, as-built drawing, specifications, previous inspection records

4. Pipe Lateral

Sources: GIS, as-built drawings, construction specifications, inspection records, other database

5. Dissimilar Materials

Sources: GIS, as-built drawings, construction specifications, inspection records, other database

6. Pipe Installation

Sources: Inspection records

7. Pipe Manufacture

Sources: As-built drawing, specifications

8. Pipe Trench Width

Sources: As-built drawing, specifications

9. Pipe external Coating

Sources: As-built drawings, construction specifications, inspection records, other database

10. Pipe Cathodic Protection

Sources: As-built drawings, construction specifications, inspection records, other database

11. Pipe Thrust Restraint

Sources: As-built drawings, construction specifications, inspection records, other database

12. Lateral Connections

Sources: GIS, as-built drawings, construction specifications, inspection records, other database

13. Pumping Station and WWTP

Sources: Utility records, GIS, plans, and construction specifications

14. Sewer Odors

Sources: Inspection records, Operational records, customer complaints

15. Sewer Flooding

Sources: Previous studies, customer complaints records, flooding record

16. Sewer Overflow (SSO/CSO)

Sources: Local sewer record plans, GIS, survey records, SSO/CSO event loggers

17. Backup floodings

Sources: Previous studies, customer complaints records, utility records

18. Leakage allowance

Sources: Local regulations, construction specifications, utility records

19. Interruptions

Sources: Operational records, customer complaints

20. Soil Disturbance

Sources: Inspection records, field inspections, operational records

21. Runoff Rate

Sources: Utility records (storm drainage), Hydraulic Model

Inspection/Testing:

- Flow monitoring
- Ruoff Rate Calculation

Ref:

- Wastewater Collection System Modeling and Design, First Edition, Heastad M. et al. , 2004

22. Non-Uniform Soil

Sources: Inspection records, field inspections, geotechnical records

- Geospatial data gateway - http://datagateway.nrcs.usda.gov/
- Natural Resources Conservation Service - http://soils.usda.gov/survey/geography/statsgo/

Inspection/Testing:

- American Society for Testing and Materials: ASTM D422 - 63(2007) Standard Test Method for Particle-Size Analysis of Soils
- American Society for Testing and Materials: ASTM D3282-93(2004)e1 Standard Practice for Classification of Soils and Soil-Aggregate Mixtures for Highway Construction Purposes
- American Association of State and Highway Transportation Officials : AASHTO M 145 Standard Specifications for Classification of Soils and Soil-Aggregate Mixtures for Highway Construction Purposes

23. Frost Penetration

Sources:

- Many state DOTs have frozen soil profiles - http://www.mrr.dot.state.mn.us/research/seasonal_load_limits/thawindex/frost_thaw_graphs.asp

24. Non-Uniform Slope

Sources: GIS, as-built drawing, specifications, previous inspection records

- U.S. Geological Survey (USGS) Landslide Hazards Monitoring and Assessment - http://geology.usgs.gov/connections/blm/hazards/landslide_assess.htm

25. Unstable Slope

Sources: GIS, inspection records, field inspections, other databases

- U.S. Geological Survey (USGS) Landslide Hazards Monitoring and Assessment - http://geology.usgs.gov/connections/blm/hazards/landslide_assess.htm

26. Seismic Activity

Sources: Seismograph

- U.S. Geological Survey (USGS) -http://earthquake.usgs.gov/
- Incorporated Research Institutions for Seismology (IRIS) -http://www.iris.edu/

27. Catchment Area (Sewershed)

Sources: Local sewer record plans, GIS

28. Average Closeness to Trees

Sources: Derived from imagery

29. Tidal Influences

Sources: Utility Records

30. Soil pH

Inspection/Testing:

- AASHTO T 289-91 Standard Method of Test for Determining pH of Soil for Use in Corrosion Testing
- ASTM G51
- pH Meter-A rod which is inserted into moistened soil and measures the concentration of hydrogen ions electronically.
- ph Test Based on Barium Sulfate
- Litmus Paper-A sample of soil is mixed with distilled water, into which a strip of litmus paper is inserted. If the soil is acidic the paper turns read, if alkaline, blue.

31. Soil Chloride

Inspection/Testing:

- AASHTO T 291-91 I Standard Method of Test for Determining Water-Soluble Chloride Ion Content in Soil
- ASTM D512

32. Soil Sulfate

Inspection/Testing:

- AASHTO T 290-91 I Standard Method of Test for Determining Water-Soluble Sulfate Ion Content in Soil
- ASTM D516

33. Soil Sulfide

Inspection/Testing:

- Sulfide

Ref:

- Sulfide and Corrosion prediction and Control, American Concrete Pipe Association, 1998

34. Pipe Connections

Sources: Local zoning maps, operational records, other databases

35. Annual Maintenance Cost

Sources: Utility Financial records

36. Annual Repair/Rehabilitation Cost

Sources: Utility Financial records

37. Installed and Replacement Cost

Sources: Utility Financial records

38. Annual Operational Cost

Sources: Utility Financial records

39. Annual Energy Cost

Sources: Utility Financial records

40. Depreciated Value

Sources: Utility Financial records

41. Benefit/Cost

Sources: Utility Financial records

42. Density of Connections

Sources: GIS, operational records, inspection records, as build plans and specifications

43. Resident Population Served

Sources: GIS, customer billing information, census data, previous studies or researches

44. Failing Utilities

Sources: Failure records, operational records, inspection records, other databases such as transportation, water, electricity databases

45. Consequence/Risk

Sources: Utility records

Inspection/Testing:

- Evaluation of Risk
- Risk Cost
- Failure probability versus consequence

Ref:

- Protocols for Assessing Condition and Performance of Water and Wastewater Assets, WERF Condition Assessment Protocols (CAP) Project, Urquhart, T.,2006
- Decision-Support Tools for Predicting the Performance of Water Distribution and Wastewater Collection Systems, Stone S. et al., US EPA

46. Third Party Damage

Sources: Failure records, operational records, inspection records, other databases

47. Other Information

DATA COLLECTION METHODS AND PROTOCOLS REFERENCE

Birks, A., and Green, R., "Nondestructive Testing Handbook," 2nd ed., vol. 7, American Society for Nondestructive Testing, Columbus, OH, 1991.

Makar, J. M., "Diagnostic Techniques for Sewer Systems," Journal of Infrastructure Systems, 1999.

Allouche, E., and Freure, P., "Management and Maintenance Practices of Storm and Sanitary Sewers in Canadian Municipalities," The University of Western Ontario, Geotechnical Research Centre, Department of Civil and Environmental Engineering ICLR Research Paper Series – No. 18, April 2002.

Jason Consultants LLC, "Inspection Guidelines for Ferrous and Non-Ferrous Force Mains," Project 04-CTS-6UR/6aUR, Water Environment Research Foundation (WERF), Quarterly Progress Report 1, 2007.

Environmental Protection Agency (EPA), "Evaluation of Opportunities to Improve Structural Integrity Monitoring (SIM) Capabilities for Water Mains through Federal Research and Technology Transfer," EPA Report Number: CWX089, 2006.

Sinha, S. K., "A Multi-Sensory Approach to Structural Health Monitoring of Buried Sewer Pipelines Infrastructure System," American Society of Civil Engineers (ASCE), 2004.

Urquhart, T., et al., "Protocols for Assessing Condition and Performance of Water and Wastewater Assets, WERF Condition Assessment Protocols (CAP) Project," Draft Final Report WERF Project 03-CTS-20CO, 2006.

American Society of Civil Engineers (ASCE), "Manual Manhole Inspection and Rehabilitation," Engineering Practice No. 92, USA.

Heastad, M., Walski, M. T., Barnard, E. T., Harold, E., Merritt, B. L., Walker, N., Whitman, E. B., "Wastewater Collection System Modeling and Design," First Edition, Heastad Press, Waterbury CT, 2004.

ASCE/WEF, "Existing Sewer Evaluation and Rehabilitation," ASCE Manual No.62 or WEF MOP FD-6, Edition 2, 1994.

Wastewater Collection Systems Management, Manual of Practice No. 7, Water Environment Federation, 5th Edition, 1999.

U.S. Environmental Protection Agency (US EPA), "Sewer System Infrastructure Analysis and Rehabilitation handbook," EPA 625691030, 1991.

NASSCO, "Pipeline Assessment and Certification Program (PACP) Reference Manual," Pikeville, MD, 2003.

EPA Office of Water Washington, D.C., "Collection Systems O&M Fact Sheet Sewer Cleaning and Inspection," EPA 832-F-99-031, September 1999.

ANSI/AWWA C105/A21.5, "American National Standard for Polyethylene Encasement for Ductile-Iron Pipe Systems," American Water Works Association, Denver, Colorado, 1999.

Najjaran, H., Sadiq, R., Rajani, B., "Fuzzy Expert System to Assess Corrosion of Cast/Ductile Iron Pipes from Backfill Properties," Computer–Aided Civil and Infrastructure Engineering, 21, 1, pp. 67-77, 2006.

Bonds W. R, "Stray Current Effects on Ductile Iron Pipe, "Ductile Iron Pipe Research Association (DIPRA), 1997.

American Concrete Pipe Association, "Concrete Pipe Design Manual," 2000.

Frederick R. M., "Design Manual: Sulfide and Corrosion prediction and Control," American Concrete Pipe Association, ISBN 0-960-38689-0, TX, 1998.

American Concrete Pipe Association "Concrete Pipe Handbook", ISBN 0-960-38681-6, TX, 1998.

Ductile Iron Pipe Research Association (DIPRA), "Design of Ductile Iron Pipe", DIPRA/6-03/5M, AL, 2003.

Clay Pipe Institute, "Clay Pipe Handbook," WI, 2004.

Sikora J. E., "Load Factors and Non-Destructive Testing of Clay Pipe," Reprinted from Journal Water Pollution Control, 1980.

Stone S. et al., "Decision-Support Tools for Predicting the Performance of Water Distribution and Wastewater Collection," US EPA 600/R-02/029.

REFERENCES

[1] Water Infrastructure Network, *"Clean and Safe Water for the 21st Century,"* Water Infrastructure Network, Washington, D.C., 2000.

[2] United States Environmental Protection Agency (U.S. EPA), *"Managing for Excellence: Analysis of Water and Wastewater Utility Management Systems,"* Washington, D.C., 2005.

[3] American Society of Civil Engineers (ASCE), *"Report Card for America's Infrastructure,"* Reston, VA, 2005.

[4] American Society of Civil Engineers (ASCE), *"Clean Water Infrastructure Financing,"* Policy Statement 480, Reston, VA, 2000.

[5] American Society of Civil Engineers (ASCE), *"Wastewater Facilities Construction Funding,"* Policy Statement 326, Reston, VA, 2000.

[6] Grigg, N.S., *"Water, Wastewater, and Stormwater Infrastructure Management,"* CRC, ISBN 1566705738, 2002.

[7] Heastad, M., Walski, M.T., Barnard, E.T., Harold, E., Merritt, B L., Walker, N., and Whitman, E.B., *"Wastewater Collection System Modeling and Design,"* First edition, Heastad Press, Waterbury CT, 2004.

[8] Najafi, M., *"Trenchless Technology-Pipeline and Utility Design, Construction, and Renewal,"* Water Environment Federation (WEF Press), Alexandria, VA, 2005.

[9] Metcalf & Eddy, Inc. (revised by Tchobanoglous, G. and Burton, F. L.), *"Wastewater Engineering: Treatment, Disposal, and Reuse,"* 3rd edition, McGraw-Hill, NY, 1991.

[10] U.S. Environmental Protection Agency, *"Water Recycling and Reuse: The Environmental Benefits,"* IX - EPA 909-F-98-001, U.S. EPA Water Division Region, 1998.

[11] U.S. Environmental Protection Agency, *"Guidelines for Water Reuse,"* EPA/625/R-92/004, U.S. EPA Office of Technology Transfer and Regulatory Support, Sep., 1992.

[12] Schrock, B.J., "*Pipeline Systems Rehabilitation Workshop*," San Jose, CA, 1991.

[13] Makar, J.M., *"A Preliminary Analysis of Failures in Grey Cast Iron Water Pipes,"* Engineering Failure Analysis, Vol. 7, pp. 43-53, 2000.

[14] Rajani, B.B., Zhan, C., and Kuraoka, S., *"Pipe-Soil Interaction Analysis of Jointed Water Mains," Can. Geotechn. J.* 33, no. 3, pp. 3-11, 1996.

[15] Boot, J.C., *"Elastic Buckling of Cylindrical Pipe Linings With Small Imperfections Subject to External Pressure,"* Journal of Intl. Society of Trenchless Technology, vol. 12, no. 2, 1998.

[16] Heger, F.J., Liepins, A.A., and Selig, E.J., "SPIDA: An Analysis and Design for Buried Concrete Pipe," *Proceedings of the Intl. Conference of Pipelines*, American Society of Civil Engineers, Herndon, VA, pp. 143-154, 1985.

[17] Davies, J.P., Whiter, J.J., Clark, B.A., Ockleston, G.O., and Cunningham, R.J., "Application of Interaction Matrices to the Problem of Sewer Collapse," *Proceedings of 11th European Sewage and Response Symposium*, Munich, 1999.

[18] Serpente, R.F., "Understanding the Models of Failure for Sewers," *Proceedings of Intl. Conference of Pipeline Division*, American Society of Civil Engineers, New York, pp. 86-100, 1993.

[19] United States Environmental Protection Agency (U.S. EPA), *"Preamble of the Draft Proposed SSO Rule,"* U.S. Environmental Protection Agency, Washington, D.C., 2001.

[20] Black & Veatch, *"Optimization of Collection System Maintenance Frequencies and System Performance,"* ASCE, EPA Cooperative Agreement #CX 824902-01-0, U.S. Environmental Protection Agency, Washington, D.C., 1999.

[21] Lai, F., *"Review of Sewer Design Criteria and RDII Prediction Methods,"* U.S. EPA/600/R-08/010, U.S. Environmental Protection Agency, Washington, D.C., January 2008.

[22] United States Environmental Protection Agency. *"The Clean Water and Drinking Water Infrastructure Gap Analysis,"* EPA-816-R-02-020, Office of Water (4606M), U.S. Environmental Protection Agency, Washington, D.C., September 2002.

[23] Urban Institute, *"Capital infrastructure investment needs and financing options,"* HUD-0003404, Department of Housing and Urban Development, Washington, D.C., 1981.

[24] Tafuri, N.A. and Selvakumar, A., *"Wastewater collection system infrastructure research needs in the USA,"* U.S. Environmental Protection Agency, Urban Water 4 (2002) 21-29, Edison, NJ, 2001.

[25] American Society of Civil Engineers, *"State and local public works needs,"* Civil Engineering Research Foundation, December, U.S. Environmental Protection Agency, Washington, D.C., 1994.

[26] Reed, C., Robinson, J.A., and Smart, D., *"Potential Techniques for the Assessment of Joints in Water Distribution Pipelines,"* American Water Works Association Research Foundation, Denver, CO, 2006.

[27] U.S. Environmental Protection Agency, *"Wastewater Technology Fact Sheet Pipe Construction and Materials,"* EPA 832-F-00-068, Office of Water, U.S. Environmental Protection Agency, Washington, D.C., September 2000.

[28] American Concrete Pipe Association, *"Concrete Pipe Design Manual,"* American Concrete Pipe Association, Irving, TX, 2000.

[29] Clay Pipe Institute, *Clay Pipe Handbook*, National Clay Pipe Institute, Lake Geneva, WI, 2004.

[30] Ductile Iron Pipe Research Association, *"Installation Guide for Ductile Iron Pipe,"* DIPRA, Birmingham, AL, 2003.

[31] Ductile Iron Pipe Research Association, *"Ductile Iron Pipe,"* DIP-GEN/3-03/5M; DIPRA, Birmingham, AL, 2001.

[32] The Uni-Bell PVC Pipe Association, *"Handbook of PVC Pipe Design and Construction,"* Dallas, TX, 2001.

[33] American Water Works Association, *"PE Pipe—Design and Installation Manual of Water Supply Practices M55,"* ISBN 1-58321-387-2, American Water Works Association, Denver, CO, 2005.

[34] Urquhart, T. and Burn, S., "Protocols for Assessing Condition and Performance of Water and Wastewater Assets", Draft Final Report WERF Project 03-CTS-20CO, Water Environmental Research Foundation (WERF) Condition Assessment Protocols (CAP) Project, Alexandria, VA, 2006.

[35] Computer Aided Rehabilitation of Sewer Networks (CARE-S), European Commission – Fifth Framework Program, Technical Report EVK1-CT-2002-00106, Accessed at: http://care-s.unife.it, 2002.

[36] National Association of Sewer Service Companies, *"Pipeline Assessment and Certification Program (PACP) Reference Manual,"* NASSCO, Pikesville, MD, 2003.

[37] Federation of Canadian Municipalities and National Research Council, *"Assessment and Evaluation of Storm and Wastewater Collection Systems,"* A Best Practice by the National Guide to Sustainable Municipal Infrastructure, ISBN 1-897094-72-8, National Research Council Canada, Ottawa, Canada, July 2004.

[38] Stone, S. et al., *"Decision-Support Tools for Predicting the Performance of Water Distribution and Wastewater Collection Systems,"* EPA/600/R-02/029, National Risk Management Research Laboratory Office of Research and Development, U.S. Environmental Protection Agency, Cincinnati, OH.

[39] Farshad, M., *"Plastic Pipe Systems: Failure Investigation and Diagnosis,"* ISBN 1-85617-496-4, Elsevier UK, August **2006**.

[40] Cassa, A.M., *"A numerical investigation into the behaviour of leak openings in pipes under pressure,"* Thesis, Civil and Urban Engineering, University of Johannesburg, 2006.

[41] Anonymous, *"External Corrosion and Protection of Ductile Iron Pipe,"* Retrieved November 2007, http://www.angelfire.com/pop/myfile/EXTDIPhtml.htm.

[42] Burn, S. et al., *"Long-Term Performance Prediction for PVC Pipes,"* American Water Works Association Research Foundation, Denver, CO, 2005.

[43] Garcia, C., Abraham, M.D. Gokhale, S., and Iseley, T., *"Rehabilitation Alternatives for Concrete and Brick Sewers,"* ASCE Practice Periodical on Structural Design and Construction, November 2002. ASCE, Herndon, VA, 2002.

[44] Grigg, N, *"Main Break Prediction, Prevention, and Control,"* American Water Works Association Research Foundation, Denver, CO, 2007.

[45] Kellagher, R. et al., *"Development of Guideline for the structural Hydraulic and Environmental Rehabilitation of Sewer,"* SMT4 – CT98 – 2272, http://www.hydroinformatics.org/SRguide/home.html, European Union Commission, 2002.

[46] Jason Consultants, LLC, *"Inspection Guidelines for Ferrous Force Mains,"* Interim Report Water Environmental Research Foundation, Alexandria, VA, July 2007.

[47] Kleiner, Y., Rajani, B., and Sadiq, R., *"Risk Management of Large-Diameter Water Transmission Mains,"* American Water Works Association Research Foundation, Denver, CO, 2005.

[48] Le Gouellec, Y.A., and Cornwell, D., *"Installation, Condition Assessment, and Reliability of Service Lines,"* American Water Works Association Research Foundation, Denver, CO, 2007.

[49] Makar, J.M., Desnoyers, R., and McDonald, S.E., *"Failure Modes and Mechanisms in Gray Cast Iron Pipe,"* National Research Council Canada. Presented at Underground Infrastructure Research 2001, Waterloo, Ontario, June 10-13, 2000.

[50] Moser, A.P., *"Buried Pipe Design,"* Second edition, The McGraw Hill Companies Inc., New York, 2001.

[51] National Research Council Canada, *"Deterioration and Inspection of Water Distribution Systems, A Best Practice by the National Guide to Sustainable Municipal Infrastructure,"* National Research Council Canada, Ottawa, Canada, April 2003.

[52] Garcia, C., Abraham, D.M., Gokhale, S., and Iseley T., *"Rehabilitation Alternatives for Concrete and Brick Sewers,"* Practice Periodical on Structural Design and Construction, Nov. 2002.

[53] Rothman, P.S. and Price, R.E., *"Detection and Considerations of Corrosion Problems of Prestressed Concrete Cylinder Pipe,"* American Society for Testing and Materials STP906-EB, Corrosion Effect of Stray Currents and the Techniques for Evaluating Corrosion of Rebars in Concrete, Philadelphia, PA, 1986.

[54] Reed, C., Robinson, J.A., and Smart, D., *"Techniques for Monitoring Structural Behaviour of Pipeline Systems: Subject Area: Infrastructure Reliability,"* American Water Works Association Research Foundation, Denver, CO, 2004.

[55] National Research Council Canada, *"Failures in Gray Cast Iron Distribution Pipes,"* Available at: http://irc.nrc-cnrc.gc.ca/ui/bu/grayindex_e.html.

[56] Bonds, W.R, *"Stray Current Effects on Ductile Iron Pipe,"* Ductile Iron Pipe Research Association, Birmingham, AL, 1997.

[57] Wienhold, P. and Chudnovsky, A., *"Lifetime of High-Density Polyethylene Drain Pipe in an Aggressive Environment,"* Transportation Research Board Annual Meeting 2006 Paper #06-2377, Washington, D.C., 2006.

[58] Rajani, B. et al., *"Investigation of Grey Cast Iron Water Mains to Develop a Methodology for Estimating Service Life,"* American Water Works Association Research Foundation, 2000.

[59] Starling, N. et al., *"Characterization of Lead Pipes Used in the Water Industry: Extrusion Processing, Alloy Microstructure and Their Role in Service Failures,"* Materials Science and Technology, vol. 23, no. 5, pp. 600-605, May 2007.

[60] Szeliga, M.J. and Simpson, D.M., *"Evaluating Ductile Iron Pipe Corrosion,"* Pipelines 2003.

[61] UK Water Industry Research (UKWIR), *"Nationally Agreed Failure Data and Analysis Methodology for Water Mains: Volume II,"* Protocol for the Capture and Reporting of Data on Mains Failures, UK, 2002.

[62] Al-Barqawi, H. and Zayed, T., *"Condition Rating Model for Underground Infrastructure Sustainable Water Mains,"* ASCE Journal of Performance of Constructed Facilities, vol. 20, no.2, pp. 126-135, May 2006.

[63] Computer-Aided Rehabilitation of Water Networks, *Construction of a Control Panel of Performance Indicators for Rehabilitation, Report #2 Validation of the Rehabilitation Performance Indicators System,* Lisbon, March 2002, pp. 1-95.

[64] Chughtai, F. and Zayed, T., "Structural Condition Models for Sewer Pipeline," Pipelines, *ASCE Conference Proceedings*, 2007.

[65] Deb, A.K., Grablutz, F.M., Hasit, Y.J., Snyder, J.K., Loganathan, G.V., and Agbenowski, N, *"Prioritizing Water Main Replacement and Rehabilitation,"* American Water Work Association Research Foundation, Denver, CO, 2002.

[66] Elsayegh, H.K., Hutchinson, R.E., and Norris, L., *"Clean Water Atlanta: A Comprehensive approach to QA/QC from SSES to Rehabilitation Design"*, Water Environment Federation, Alexandria, VA, 2007.

[67] U.S. Environmental Protection Agency, *"Collection Systems O&M Fact Sheet Sewer Cleaning and Inspection,"* EPA 832-F-99-031, Office of Water Washington, D.C., September 1999.

[68] Kathula, V., *"Structural Distress Condition Modeling for Sanitary Sewer,"* Ph.D. Thesis, Civil Engineering, Louisiana Technological University, 2001.

[69] Mays, L.W., *"Urban Water Supply Handbook,"* McGraw Hill Companies Inc., New York, 2002.

[70] Mehta, C.R., *"Identification and Characterization of Parameters for Sewer Pipe Condition Rating,"* MS Thesis, Pennsylvania State University, PA, 2006.

[71] Nelson, R.E., Hsiung, P.H., and Witt, A.A., *"Optimization of Collection System Maintenance Frequencies and System Performance,"* American Society of Civil Engineers, Washington, D.C., 146, 1999.

[72] Royer, M.D., *"White Paper on Improvement of Structural Integrity Monitoring for Drinking Water Mains,"* EPA/600/R-05/038, U.S. Environmental Protection Agency, Washington, D.C., 2005.

[73] Birks, A. and Green, R., *"Nondestructive Testing Handbook,"* 2nd ed., vol. 7, American Society for Nondestructive Testing, Columbus, OH, 1991.

[74] Makar, J. M., "Diagnostic Techniques for Sewer Systems," *J. Infrastruct. Syst.,* 1999.

[75] Allouche, E. and Freure, P., *"Management and Maintenance Practices of Storm and Sanitary Sewers in Canadian Municipalities,"* The University of Western Ontario, Geotechnical Research Centre, Department of Civil and Environmental Engineering ICLR Research Paper Series – No. 18, April 2002.

[76] Water Environment Research Foundation, *"Inspection Guidelines for Ferrous and Non-Ferrous Force Mains,"* Project 04-CTS-6UR/6aUR, Quarterly Progress Report 1, 2007.

[77] Sinha, S.K., *"A Multi-Sensory Approach to Structural Health Monitoring of Buried Sewer Pipelines Infrastructure System,"* American Society of Civil Engineers, New York, 2004.

[78] U.S. Environmental Protection Agency, *"Evaluation of Opportunities to Improve Structural Integrity Monitoring (SIM) Capabilities for Water Mains through Federal Research and*

Technology Transfer," EPA Report Number: CWX089, U.S. Environmental Protection Agency, Washington, D.C., 2006.

[79] American Society of Civil Engineers, *"Manual Manhole Inspection and Rehabilitation,"* Engineering Practice No. 92, New York, ASCE, 2009.

[80] American Society of Civil Engineers/Water Environment Federation, *"Existing Sewer Evaluation and Rehabilitation,"* ASCE Manual No.62, New York, ASCE, 1994.

[81] Water Environment Federation, *"Wastewater Collection Systems Management,"* Manual of Practice No. 7, 5th edition, Water Environment Federation, Alexandria, VA, 1999.

[82] U.S. Environmental Protection Agency, *"Sewer System Infrastructure Analysis and Rehabilitation Handbook,"* EPA 625691030, U.S. Environmental Protection Agency, Washington, D.C., 1991.

[83] American National Standard Institute/American Water Work Association, *"American National Standard for Polyethylene Encasement for Ductile-Iron Pipe Systems,"* C105/A21.5, American Water Works Association, Denver, CO, 1999.

[84] Najjaran, H., Sadiq, R., and Rajani, B., *"Fuzzy Expert System to Assess Corrosion of Cast/Ductile Iron Pipes from Backfill Properties,"* Computer-Aided Civil and Infrastructure Engineering, 21(1), 67-77, 2006.

[85] Frederick, R.M., *"Design Manual: Sulfide and Corrosion prediction and Control,"* American Concrete Pipe Association, Irving, TX, 1998.

[86] American Concrete Pipe Association *"Concrete Pipe Handbook,"* Author: Irving, Tex., 1998.

[87] Ductile Iron Pipe Research Association, *"Design of Ductile Iron Pipe,"* DIPRA/6-03/5M, Birmingham, AL, 2003.

[88] Sikora, J.E., "Load Factors and Non-Destructive Testing of Clay Pipe," Reprinted from *J. Water Pollut. Control,* Water Environment Federation, Alexandria, VA, 1980.

[89] Hutchinson, E.R., El-Sayegh, K.H., and Chambers, L., *"Atlanta Track – Paper Two Atlanta's SSES & Integrated Sewer Rehabilitation Selection Process,"* Pipelines 2007: Advances and Experiences with Trenchless Pipeline Projects, American Society for Civil Engineers, Herndon, VA, 2007.

[90] City of Atlanta Department of Watershed Management, *"Asset Management Case Study,"* Available at: https://courses.worldcampus.psu.edu/public/buried_assets/resources/atlanta.pdf.

[91] Pittsburgh Water and Sewer Authority, *"Pittsburgh Water and Sewer Authority – Homepage,"* Available at: http://www.pgh2o.com/.

[92] Pittsburgh Water and Sewer Authority, *"Monitoring & Sampling Plan,"* PWSA CSO Long-Term Control Plan Development Project. Available at: https://courses.worldcampus.psu.edu/public /buried_assets/resources/pittsburgh.pdf.

[93] Ross & Associates Environmental Consulting, Ltd., *"Managing for Excellence: Analysis of Water and Wastewater Utility Management Systems,"* EP-W-04-023, U.S. Environmental Protection Agency, Washington, D.C., August 2005.

[94] Brown and Caldwell, *"Seattle Public Utilities 2006 Wastewater Systems Plan,"* Seattle Public Utilities, Seattle, WA, March 2006.

[95] Martin, T., Johnson, D., and Anschell, S., "*Using Historical Repair Data to Create Customized Predictive Failure Curves for Sewer Pipe Risk Modeling*," LESAM 2nd Leading Edge Conference on Strategic Asset Management, Lisbon, Portugal, October 2007.

[96] Martin, T., "*Seattle Public Utilities Risk Management at SPU*," PowerPoint Presentation, Seattle Public Utilities, 2008.

[97] Orange County Sanitation District, *"Orange County Sanitation District Webpage,"* Available at http://www.ocsd.com/about/general_information/default.asp.

[98] Orange County Sanitation District, *"Asset Management Plan 2006,"* Available at: http://www.ocsd.com/civica/filebank/blobdload.asp?BlobID=5818.

[99] Wiley & Wilson, *"Town of Blacksburg Sanitary Sewer System Study,"* VA, 2006.

[100] Shah, A., Tighe, S., and Stewart, A., "Development of a Unique Deterioration Index, Prioritization Methodology, and Foreign Object Damage Evaluation Models for Canadian Airfield Pavement Management," *Can. J. Civil Engin.,* vol. 31, pp. 608-618, 2004.

[101] Mauch, M., and Madanat, S., *"Stochastic Duration Models of Bridge Deck Deterioration,"* 79th Meeting, Transportation Research Board, Washington, D.C., 2000.

[102] Butt, A.A., Shahin, M.Y., Feighan, K.J., and Carpenter, S.H., *"Pavement Prediction Model Using the Markov Process,"* Transportation Research Record 1123, Transportation Research Board – National Research Council, National Academy Press, Washington, D.C., pp.12-19, 1987.

[103] Shamir, U. and Howard, D., "An Analytic Approach to Scheduling Pipe Replacement," *AWWA J.*, May 1979, pp. 248-258.

[104] Abraham, D.M., Wiradhadikusumah, R., Short, T.J., and Shahbahrami, S., "Optimization Modeling for Sewer Network Management," *J. Constr. Engin. Mgmt.,* vol. 124, n. 5, pp. 402-410, 1998.

[105] Sinha, S.K. and Knight, M., "Development of an Intelligent System for Underground Pipeline Management," *J. Comp. Aided Civ. Infrastruct. Engin.*, vol. 19, no. 1, pp. 42-53, 2004.

[106] Wirahadikusumah, R., Abraham, D., and Iseley, T., "Challenging Issues in Modeling Deterioration of Combined Sewers," *J. Infrastruct. Sys.,* June 2001.

[107] Kathula, V., "*Structural Distress Condition Modeling for Sanitary Sewer*," Ph.D. Thesis, Civil Engineering, Louisiana Technological University, 2001.

[108] Tran, D.H., Ng, A.W.M., and Perera, B.J.C., "Neural networks deterioration models for serviceability condition of buried stormwater pipes," *Engin. Applic. Artific. Intell.*, 2007.

[109] Kleiner, Y., Rajani, B., and Sadiq, R., "*Risk Management of Large Diameter Water Transmission Mains*," American Water Works Association Research Foundation, Denver, CO, 2006.

[110] Wiradhadikusumah, R., Abraham, D.M., and Castello, J., "Markov Decision Process for Life Cycle Cost Based Sewer Rehabilitation," *J. Engin. Constr. Architect. Mgmt.*, Oxford, vol. 6, n. 4, pp. 358-370, 1999.

[111] Madanat, S.M., Mishalani, R., and Ibrahim, W.H., "Estimation of Infrastructure Transition Probabilities from Condition Rating Data," *J. Infrastruct. Sys.*, no. 2, pp.120-125, 1995.

[112] Madanat, S.M., Karlaftis, M.G., and McCarthy, P.S., "Probabilistic Infrastructure Deterioration Models with Panel Data," *J. Infrastruct. Sys.* 3, no. 1, pp. 4-9, 1997.

[113] Micevski, T., Kuczera, G., and Coombes, P. "Markov Model for Storm Water Pipe Deterioration," *J. Infrastruct. Sys.*, Jun. 2002.

[114] Baik,, H., Seok, H., Jeong, D., and Abraham, D. M., "Estimating Transition Probabilities in Markov Chain-Based Deterioration Models for Management of Wastewater Systems," *Journal of Water Resources Planning and Management*, Jan.-Feb. 2006.

[115] Ariaratnam, S.T., El-Assaly, A., and Yang, Y. "Assessment of Infrastructure Inspection Needs Using Logistic Models," *J. Infrastruct. Sys.*, June 2001.

[116] Rostum, J., Baur, R., Saegrov, S., Horold, S., and Schilling, W. "Predictive Service-Life Models for Urban Water Infrastructure Management," *Proceeding 8th International Conference on Urban Storm Drainage,* Sydney, NSW, Australia, 1999.

[117] Achim, D., Ghotb, F., and McManus, K.J., "Prediction of Water Pipe Asset Life Using Neural Networks," *J. Infrastruct. Sys.*, ASCE, March 2007.

IWA Publishing's authorised EU representative for General Product Safety Regulations is Diane D'Arras, 15 rue Duret, 75116 Paris, France, e-mail: safety@iwap.co.uk.

Printed and bound by CPI Group (UK) Ltd, Croydon, CR0 4YY
12/05/2026
02108876-0007